An Insider's Perspective of Prostate Cancer

An Insider's Perspective of Prostate Cancer

Understanding Effects, Management Options and Consequences

Robert Alexander ('Frank') Gardiner, AM, MD, MB BS, FRCS, FRACS

Emeritus Professor, University of Queensland
Affiliate Professor, University of Queensland
Honorary Consultant Urologist, Royal Brisbane & Women's Hospital
Adjunct Professor, Edith Cowan University, Perth, Western Australia

ELSEVIER

ACADEMIC PRESS
An imprint of Elsevier
elsevier.com/books-and-journals

Academic Press is an imprint of Elsevier

125 London Wall, London EC2Y 5AS, United Kingdom
525 B Street, Suite 1650, San Diego, CA 92101, United States
50 Hampshire Street, 5th Floor, Cambridge, MA 02139, United States
The Boulevard, Langford Lane, Kidlington, Oxford OX5 1GB, United Kingdom

ISBN: 978-0-443-18709-4

For Information on all Academic Press publications visit our website at https://www.elsevier.com/books-and-journals

Publisher: Mica H. Haley
Acquisitions Editor: Stacy Masucci
Editorial Project Manager: Sara Pianavilla
Production Project Manager: Omer Mukthar
Cover Designer: Vicky Pearson Esser

Typeset by Aptara, New Delhi, India

Contents

Foreword

The burden our society bears as a consequence of prostate cancer diagnosis and management ranks higher than most, if not all, other cancers. The typically chronic issues that stem from its diagnosis—whether cured or not—impact patients, families, and loved ones in a myriad of physical, psychological, and financial ways. The author of this monograph, Prof Robert "Frank" Gardiner, has long been sensitive to this wide-ranging impact. Spending his career as an academic urological oncologist, Frank has always been a strong advocate for the patient, and always looking to empower them through education around their condition and what the path forward may hold. Rather than take retirement quietly, Frank has continued this advocacy by authoring this comprehensive educational overview of prostate cancer through the eyes of an academic who has lived through both the scientific advances in the field along with the impact these have had on the health system.

As my first research mentor, from my first meeting through to the present day, Frank's gentle and benevolent nature has remained with me. His nature of always going out of his way to support others clearly shows again in the time volunteered to write this book. By harnessing his decades of expertise, Frank has been able to meld the many facets of prostate cancer management into very accessible language. While prostate cancer may seem to be a simple problem—it is just one cancer type essentially—even with the broadest look at this book you can clearly recognise that prostate cancer is far from *just* one disease. The aim of this book is to give patients and those interested in the topic a comprehensive knowledge about the many possible facets of their condition and the options to manage it, whether at first diagnosis or recurrence. As an academic radiation oncologist working almost entirely with prostate cancer, I cannot overstate the importance of information such as this. In the era of all-pervading data overload that results in a glut of patients who have incomplete or misleading information about their disease, the presence of a curated, considered, and expert opinion for the nonexpert reader remains highly relevant.

There are many outstanding chapters covering the breadth from basic science and anatomy through to many standards and emerging therapies. Although not essential, I certainly recommend reading the basic biology sections as these will give a foundation for many of the therapeutic advances. Most importantly, the treatment descriptions give a useful account of how these will impact on the patient. There is a lot to take in, so take your time, and you can easily bounce around between chapters easily to address questions as they arise. It is important that appropriate, useful, and unbiased information is being delivered which results in confidence about the path forward. I believe this book is an outstanding example of such and will benefit those just diagnosed with prostate cancer, and their loved ones, right through to those with progressive disease.

Professor Scott Williams
Department of Radiation Oncology,
Peter MacCallum Cancer Centre and
Sir Peter MacCallum
Department of Oncology,
University of Melbourne
March 2022

Preface

Prologue/What this book is about—and not about

Problems are part and parcel of life and, no matter how much we try to avoid them, they are inescapable so the real issue when we stumble across them (or they confront us) is how can we manage them to the best of our advantage or, how can we best get out of the pickle they cause. Interestingly, so often when considered in retrospect, many of these problems seem minor but others are not and require careful consideration so that the best possible strategies can be developed to contend with them. Being told of a diagnosis of prostate cancer is in the latter category: deciding how to proceed should be made only after the significance of the diagnosis and implications of various treatment options are fully understood. This is the starting point for this book, namely just *after* a diagnosis of prostate cancer has been made.

When told for the first time of a diagnosis of cancer the news is, at the very least, confronting. Consequently, it is far from unusual for a flood of thoughts and emotions to follow and cloud thinking at this time so that qualifying comments, management choices, and reassurances detailed by health professionals cannot be fully comprehended, let alone thoroughly appreciated. The presence of a close family member or friend at such times can be invaluable as can further consultations with the specialist who conveyed the edict or, if preferred, another doctor. Getting a second opinion is not "letting the first doctor down" or "doing the wrong thing." On the contrary, second opinions are well accepted in the medical profession and are certainly not discouraged: nobody has a monopoly on wisdom and many people relate better to some people than to others.

This book *does not* address the important topic of psychological strategies for dealing with the diagnosis and the prostate cancer journey, although it does touch on them in a number of places. Instead, the reader is referred to **"Facing the Tiger"** (second edition) by Professor Suzanne Chambers AO. This book is an invaluable survivorship guide for men with prostate cancer and their partners in dealing with the confronting news of a prostate cancer diagnosis, its implications, and the likely psychological consequences of various management options. A subsequent book by the same author, Dr Nicole Heneka and Professor Jeff Dunn AO **"The Health Professionals Guide to Delivering Psychological Care for Men with Prostate Cancer"** provides greater detail and is also strongly recommended.

In addition, this book is not designed to guide or direct an individual patient's course in choosing treatment during his prostate cancer journey. In being neither prescriptive nor proscriptive you may very well ask, what *does* this book aim to do? Its goal is to explain what happens "on the inside" physically with prostate cancer and in relation to various treatments. In doing so, I have attempted to steer a course that is not overly complex but, at the same time, is intended to provide an understandable explanation of the changes that can result from prostate cancers and from management interventions. However, there may be segments of text or even chapters that are too detailed, especially for readers not versed in biological or medical science. If so, these parts should be able to be skipped without losing the gist of the message contained in that chapter or section. On the other hand, for those wanting more information, an extensive list of references is provided.

Curricula in medical schools contain limited urological content so there exists a real risk that many doctors graduate with a paucity of knowledge of urology and prostate cancer in particular. By pursuing a strategy of not "trying to be all things to all people" and focussing on content familiar to those with a professional or scientific level of understanding and with a particular interest in the topic, it is hoped that this book will lead to improved management of prostate cancer patients.

In the not-so-distant past, a cancer diagnosis was a death sentence with demise often protracted and unpleasant. Increasingly more and more people diagnosed are living longer and surviving well. This is particularly so with prostate cancer patients due to earlier and more discriminating diagnosis pathways and better treatments as indicated subsequently.

Traditionally, prostate cancer was considered to be a disease of "old men." As such, it was generally, although erroneously, stated that "men *never* died *from* prostate cancer, they died from other conditions *with* prostate cancer" Consequently, treatment was invariably conservative and directed towards palliation and management of debilitating and painful sequelae as they presented. Diagnosis from histopathology of a biopsy was generally made after palpating a rock-hard and nodular prostate on the digital rectal exam or by symptoms and signs of primary cancerous enlargement causing urinary obstruction or secondary tumours (**metastases or "secondaries"**), such as back pain, nerve root or, least commonly, spinal cord compression. In a large majority of cases, tumours had already spread widely at the time of diagnosis and, therefore, were incurable. It was in the mid-1980s, with the introduction of the prostate-specific antigen blood test that prostate cancer began to be diagnosed earlier when it was curable and was detected increasingly at a younger age.

Advanced and aggressive cancers have a repertoire of changes they are able to employ to ensure their survival and advancement making prostate cancer potentially such a formidable adversary. Many of these adaptations are shared with other cancers but others are prostate cancer-specific.

In appreciating the many and varied changes that can and do occur in the minority of men with really aggressive tumours, it is interesting to consider from the cancer's perspective the potential options that it may usurp and pervert to proceed. Therefore, imagine that you were the aggressive and ruthless tumour, hell-bent on thriving at any cost and prepared to exploit every aspect of normal bodily functioning to achieve this end including destroying the host.

Fortunately for patients, this degree of "tumour megalomania" is present in only a minority of instances as most prostate cancers are not inherently so aggressive nor do they get the chance to orchestrate many of these changes directed to aiding their progression. In appreciating at least some of the changes that can occur, I trust that you find the contents both engaging and interesting as well as helpful in understanding this extremely common condition.

RAG Brisbane November 2021

Acknowledgements & dedication

This book is dedicated to all men diagnosed with prostate cancer, their loved ones, and those who care for them. It is a personal interpretation resulting from a culmination of many years of training, practicing, and continuous learning as a urologist and researcher. I am most grateful to have been the recipient of much generosity and guidance from many colleagues, both here and overseas. Among my urological mentors, two, in particular, the late Dr Alex Splatt in Brisbane and late Mr John Wickham FRCS Esq in London, warrant particular mention for their unflagging encouragement and support over many years.

In addition, I am indebted to a number of friends and colleagues from a wide range of backgrounds who critiqued the text and provided constructive feedback. Their opinions and advice have been invaluable in terms of providing a range of suggestions, correcting misapprehensions, and helping with readership targeting. Thus, I am indebted to Dr Michael Bolton, Peter Dornan AM, Professor Suzanne Chambers AO, Professor Jeff Dunn AO, Dr Kevin Koo, Emeritus Professor Martin Lavin, Professor Rob McLachlan AM, Leanne Morton, Emeritus Professor Teik Oh AM, Dr Matthew Roberts, Professor Hema Samaratunga, Brian Trevethan, and especially Associate Professor Michelle Hill, Associate Professor John Yaxley, and Professors Rob Newton and Scott Williams. Professor Freddie Hamdy, Editor of BJU International, kindly gave permission to use illustrations from FingerTip Urology, which was a freely available educational resource available web-based resource accessible via the BJUI website.

The contents were initially planned to target educated readers seeking more than just the superficial level of information provided for the general public, presented at a primary school literacy level and available free through many organisations. Consequently, I have tried to pitch the text comparably to that used in scientific and medical articles in *The Economist* news magazine which caters to a global audience of discerning readers. However, as writing proceeded, I became increasingly aware of the need for an up-to-date, highly referenced "one-stop-shop" that would also appeal to medical and science graduates, both recent and not so recent, as well as prostate cancer researchers and readers without a medical background.

Being both comprehensive and current in such a book is at best an ambitious and temporary phenomenon. Although many established treatments have stood the test of time and explanations of their modes of action—the main theme of the book—will not change, further details of how other interventions work will become clearer through ongoing research leading to increased options for diagnosis, monitoring, and therapy.

Undertaking an exercise such as this has given me an opportunity to reflect on many things. One of these is appreciating just how fortunate I have been in the lottery of life—as if I need reminding! Not only did I choose outstanding, encouraging, and loving parents, but I have been exceptionally fortunate with my family, and especially my better half Roslyn, to whom I am particularly indebted, who have provided unwavering support and encouragement in this and all my endeavours.

RAG November 2021

Chapter 1

Introduction

1.1 The spectrum of prostate cancers

Prostate cancer is prevalent throughout the developed world. In Australia, it is the most common cancer in men [1,2] and the second most common cause of cancer deaths in males after lung cancer [2]. More men die from prostate cancer than do women from breast cancer [3]. With few exceptions, it is a condition of older men with few diagnosed before the age of 45 years but, unlike most other malignancies, the large majority of prostate cancers progress relatively slowly. Consequently, in terms of influencing prognosis, a few days, weeks, or even months are not critical for an individual to decide the most appropriate form of treatment. Although this statement does not endorse endless procrastination in decision-making, there is no need to be "pressured into" a rushed decision. Because of the overall slow progression rate, it is the 10-year and 15-year follow-up, rather than at 5 years as with most cancers, that is more appropriate for indicating the success or otherwise of various treatments directed at the cure for the majority of men who present with clinically localised disease.

After the age of 45, the incidence of prostate cancer becomes increasingly common with advancing years. This tumour tends to run in families: men with a family history of breast or prostate cancer have a greater likelihood of developing the disease: those with a family history of prostate cancer [4] have been reported to be diagnosed approximately 6 years earlier, particularly if they harbour germline mutations (heritable DNA changes in a cell destined to become an egg or a sperm) in the family susceptibility genes BRCA1 and BRCA2. Variants in BRCA mutation carriers are thought to convey different risks for aggressive prostate cancer developing [5], which depend on the location(s) of the specific inherited mutation(s) [6,7]. BRCA2 mutations have been reported to be uncommon in sporadic prostate cancers having been estimated to be $<1\%$ in US patients and 2.3% in UK patients [5]. In addition, nongermline or somatic mutations can develop. These alterations can be spontaneous with contributions and interactions from both endogenous and exogenous factors.

It is said that if men live long enough, all will develop prostate cancer: but many cancers are of a low malignant form and do not progress so do not require treatment. If these low-risk cancers are diagnosed, all detection does is

An Insider's Perspective of Prostate Cancer: Understanding Effects, Management Options and Consequences.
DOI: https://doi.org/10.1016/B978-0-443-18709-4.00009-2

to invoke worry and concern in the minds of those in whom they are found, not to mention their loved-ones. These low-grade cancers were diagnosed more often in the fairly recent past as the very large majority of patients with an abnormally elevated serum prostate specific antigen (**PSA**) level proceeded to biopsy. In order to alleviate both doctor and patient concerns, many men with what we now know were low-risk cancers underwent treatment as there was uncertainty that even multiple, spatially positioned diagnostic biopsies may have missed a more aggressive lesion beyond the sampling needle, a concern only resolved in the past with whole gland removal and histopathological scrutiny of every skerrick of prostatic tissue.

Accusations of overdiagnosis (and overtreatment) followed with varying recommendations issued regarding the benefits or otherwise of early detection of prostate cancer along with the refuge statement by detractors that "more men die with prostate cancer than from it." However, decisions to diagnose should not be made on who will die or not die from the disease, they need to be made in the context of whether or not detection will benefit the individual patient. It is his overall welfare, both physical and otherwise, that remains the paramount determinant. Interestingly, from a health economic perspective, early detection of *clinically significant* prostate cancer is considerably cheaper with prostate cancer management costs varying with disease stage at presentation [8,9].

Selection and detection procedures for those proceeding to diagnostic prostatic biopsy have changed considerably over the past few years. These are a result of significant improvements in tumour categorisation through refinements in prostatic magnetic resonance imaging (**MRI**) and the internationally agreed prostate imaging-reporting and data system (**PIRADS**) [10]. Consequently, almost all men with an elevated PSA in whom a diagnosis of prostate cancer is considered likely to be beneficial, now proceed to a prostatic MRI. This approach of triaging with MRI decreases the number of "unnecessary" biopsy procedures of patients whose elevated PSA is not due to prostate cancer and those with low grade, clinically insignificant prostate cancer (usually PIRADS 1–2). Integration of abnormal MRI findings (PIRADS 3-5) with trans-rectal ultrasound (**TRUS**) imaging (and possibly prostate specific membrane antigen radioisotope imaging—see Staging of Prostate Cancer) also enables accurate targeting of areas of concern highlighted by MRI [10–12].

The diagnosis of prostate cancer continues to rely on the highly specialised though subjective interpretation of the histopathology submitted to uropathologists from cores of tissue obtained by needle biopsies. Although prostate biopsies are assessed for a number of pathological findings, cancer grade remains the most important prognostic feature on which treatment is based. The Gleason score/International Society of Urological Pathologists (**ISUP**) grade is regarded as the diagnostic reference with molecular profiling using combinations of mutational changes expected to have a bigger role in how prostate cancer will be managed in the future with combinations of genetic variants indicating a man's inherited risk of prostate cancer and, after cancer is diagnosed, time to biochemical recurrence and risk of metastatic disease [13].

Prostate cancers are often heterogeneous meaning that there are parts that are further in their appearance from normal prostate under the microscope with the more aggressive tumour foci referred to as being more de-differentiated (compared with normal). Because of this variation in dedifferentiation in tumour biopsies from the same prostate, a number of grading systems have been devised over the years (such as those proposed by Gaeta, Mostofi, Bocking and Gleason). However, as indicated above, the Gleason scoring system has emerged clearly as the preferred system for indicating patient prognosis. It has been refined over the years and is currently universally regarded as the histopathology prognostic reference [10].

Based on autopsy findings, in addition to low-grade and inapparent tumours, premalignant changes commonly accompany clinically significant prostate cancers. One United States study reported that of 249 cases examined, 70% of the prostates with the premalignant condition high-grade prostate intraepithelial neoplasia (**HGPIN**) harboured adenocarcinoma: the frequency of cancer in prostates without HGPIN was 24%. HGPIN was encountered in 0%, 5%, 10%, 41%, and 63% of men in the third, fourth, fifth, and seventh decades, respectively. The corresponding figures for invasive carcinoma were 2%, 29%, 32%, 55%, and 64%, respectively [14].

Although patients are classified in terms of their risk of prostate cancer progression based on clinical and investigational findings with the Gleason score of pivotal importance, the presence of other histological features such as intraductal or cribriform patterns are also associated with more aggressive forms of the disease. Combining histological and clinical parameters categorises patients as having low, intermediate of high-risk prostate cancer (see Chapter 5, Prostate Cancer Stratification). Cancers categorised as high risk when diagnosed are often more aggressive from the outset, being likely to progress especially if left untreated [15,16]. The real challenge is knowing the nature of cancer and its extent so that management can be tailored appropriately.

1.2 Diagnosis in the clinical context

In most developed countries, male life expectancy continues to increase (Table 1.1) but so does the likelihood of developing prostate cancer. It is of concern that the incidence of metastatic or secondary prostate cancer has increased in the United States over the past decade. Furthermore, in contrast to the majority of patients diagnosed with clinically localised disease, most men diagnosed with metastatic disease are reported to die from their prostate cancers [17,18] with the majority doing so within the first 2 years [17].

However, a prostate cancer diagnosis is not and should not be regarded as an automatic passport to treatment. Most patients when diagnosed have other

TABLE 1.1 Male life expectancy at birth for selected countries.

Country	Year	At birth
Switzerland	2019	81.8
Singapore	2020	81.5
Japan	2019	81.5
Australia	2018–2020	81.2
Sweden	2020	80.63
New Zealand	2017–2019	80.0
France	2019	79.7
UK	2018–2020	79.0
USA	2019	76.3

https://www.abs.gov.au/statistics/people/population/life-tables/latest-release.

medical conditions (**comorbidities**), the gravity of which they are often unaware in terms of the likelihood of these conditions leading to premature death. Among these clinical harbingers, high blood pressure (hypertension), heart and vascular disease, obesity, and diabetes (diabetes mellitus) rank highly along with, of course, a history of cigarette smoking. Many men consider themselves to be so fit that they are "bullet-proof" and likely to live indefinitely with many experienced clinicians also having difficulty in accurately predicting an individual patient's life expectancy [19].

Many validated instruments are available for indicating the risk of general comorbidity-related deaths [20], the best known of which is the **Charlson Comorbidity Index**. In addition, a form of the Total Illness Burden Index was modified and validated for men with prostate cancer (**TIBI-CaP**) and has demonstrated good utility for predicting death in a large cohort of prostate cancer patients [21]. Although these instruments are useful guides, they are not used routinely in clinical practice, being reserved mostly for selecting participants in clinical trials likely to live long enough to have a survival benefit from an intervention.

Life tables can also help in bringing us back to reality. Using Australian 2020 figures as an example (Table 1.2), ischaemic heart disease was by far the commonest cause of male deaths with almost twice the number of deaths recorded compared with the second-ranked cause, dementia. If the related category of cerebrovascular disease, the fourth commonest cause, is combined with ischaemic heart disease, this difference is even more pronounced. Prostate cancer was the sixth most common cause of male deaths with cancers of the trachea, bronchus and lung and chronic lower respiratory disease also ahead in this list. However, it is not death itself that is so much the problem, it is the process of dying and, for the minority of prostate cancer patients who develop incurable

TABLE 1.2 Leading causes of death, males, Australia 2020.

Diagnosed cause	Ranking of deaths	Number of deaths
Ischaemic heart diseases	1	10,040
Dementia (all causes)	2	5,250
Malignant neoplasms of trachea, bronchus, and lung	3	4,751
Cerebrovascular diseases	4	3,974
Chronic lower respiratory diseases	5	3,706
Malignant diseases of prostate	6	3,568

https://www.abs.gov.au/statistics/health/causes-death/causes-death-australia/2020.

disease, this can be protracted and unpleasant although it should not be and, in most cases, can be avoided (see Chapter 7: Survivorship and Palliative Care sections).

Life tables and comorbidity indices are a useful source of information but are based on populations, not individuals: not unreasonably, men want to know how likely *they* are to live and for how long. In terms of cardiovascular disease as the cause of men's demise, erectile dysfunction (**ED**) is increasingly being recognised as a sentinel event for cardiovascular morbidity and sudden death [22,23] with the median time interval between manifestation of ED and cardiovascular death 10 years in one large study [23], relevant with the10–15-year life expectancy generally required for a durable survival benefit to be realised from the time of diagnosis and treatment following a diagnosis of prostate cancer.

If treatments did not have significant side effects, all men with any form of prostate cancer would be treated definitively but unfortunately, many men do have treatment-associated problems which, on occasions, can be seriously incapacitating. Other men may have personal reasons for not wanting treatment until symptoms occur, especially if they are feeling well and enjoying life. Because of the low rate of progression of most prostate cancers, the fact that all treatments are associated with unwanted effects and life-expectancy limitations due to comorbidities and age may be present, many patients decide not to even proceed to diagnosis in addition to a number of those diagnosed choosing not to receive active treatment.

It is the whole patient and not just cancer that is to be managed with the mantra that the (effects of) treatment should not be worse than (the effects of) the disease. Prostate cancer can exact a considerable toll as can treatment associated comorbidities, so management regimens should be integrated to minimise unwanted effects and maximise the priorities and overall wishes of each patient. As a result, it is of paramount importance, that a decision made to proceed is an informed one by the patient himself with appropriate guidance from his clinician—not from Dr Google or based on what the boys say at the bar—and that the decision is made without coercion.

As stated before, because most prostate cancers progress slowly, there is seldom any need to rush this management decision as long as procrastination does not lead to endless delay. Patients need to "get on with life" after deciding their preferred courses of action.

It is hoped that this book will enfranchise readers so that they might be in a better position to understand prostate cancer and deal with issues arising from the disease and its treatments. However, before discussing prostate cancer, it is apposite to know a little background about the prostate itself, why cancers occur and how some can spread outside the gland.

Chapter 2

The "lie of the land"

2.1 Origin of the prostate

The word "prostate," derived from the Greek prohistani which means "to stand in front of" has been attributed to Herophilus of Alexandria [24]. The prostate in men is a pelvic structure situated immediately below the bladder, with which it is intimately related, and in front of the rectum (Fig. 2.1). The urethra traverses the prostate and finally the penis. The primary function of the prostate gland is to provide enzymes to maintain the fluid nature of seminal fluid and to nourish sperm as they pass through the prostatic urethra and penile urethra to the outside world.

Embryologically, the prostate develops from outgrowths of epithelial (or lining) cells from the destined prostatic segment of the urethra that grow into surrounding mesenchyme (embryonic connective tissue). Branching starts to appear in this outgrowth at approximately 10 weeks and within a couple of weeks tubules appear [25]. Embryonic growth involves the activation of many genes that function for a short time before becoming inactive after fulfilling their roles in normal development. However, reactivation of these genes may occur in later life with prostate cancer development and progression.

2.2 Anatomical considerations

The prostate gland increases in size at puberty coinciding with a surge in circulating testosterone so that in adults, the normal gland is about the size of a walnut, and, although firm and smooth to palpation via the rectum, is clearly not of the same consistency. Its depictions in diagrams and sketches imply that the prostate is a fixed pelvic structure and that it is immobile, but this is far from the case. Although attached to other pelvic structures by multiple layers of fascia (connective tissue), the prostate contracts actively during ejaculation (due to activation of its abundant smooth muscle) with its exocrine products (prostatic fluid) expelled to mix in the urethra with sperm discharged from the seminal vesicles and vas deferens via the common ejaculatory duct on each side: the prostate also moves when prodded (e.g., by a biopsy needle).

An Insider's Perspective of Prostate Cancer: Understanding Effects, Management Options and Consequences.
DOI: https://doi.org/10.1016/B978-0-443-18709-4.00007-9

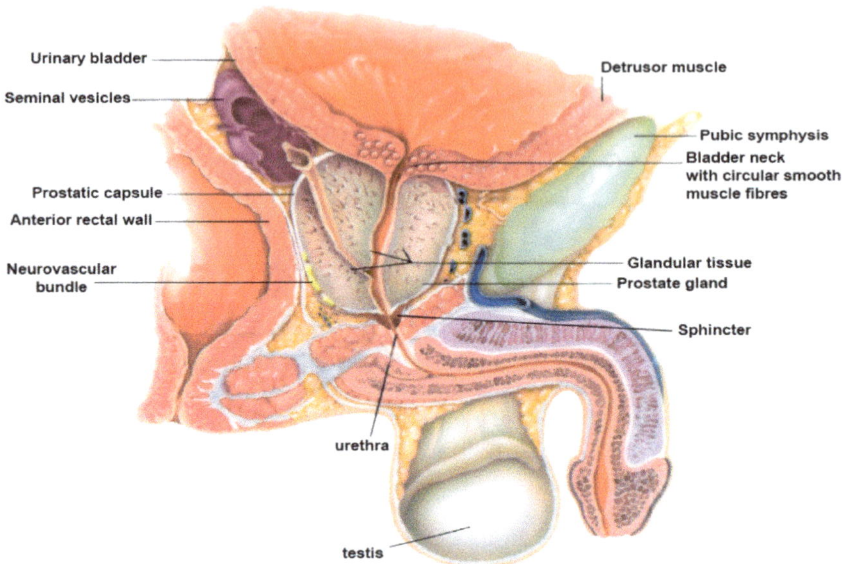

FIGURE 2.1 The normal prostate & its relationship to other pelvic structures.

The urethra traverses the prostate into which the ejaculatory ducts open and through which stored sperm pass to mix with prostatic fluid prior to being actively propelled down the urethra during ejaculation. At the upper end of the urethra is a condensation of circular smooth muscle which keeps the bladder neck closed at all times, except when the bladder wall (**detrusor**) muscle contracts, increasing the pressure in the bladder and, at the same time, widening and funnelling the bladder neck to permit urine to enter the prostatic urethra. When the detrusor is not contracting, the closed bladder neck functions as a cough-proof, strain-proof, and water-tight sphincter.

The distal (meaning further away) rhabdo-sphincter, consisting of specialised skeletal muscle with an abundance of slow-twitch fibres, is at the bottom of the prostate and overlaps its lower part to a variable degree. The rhabdo-sphincter relaxes during voiding, just prior to bladder-neck opening, to enable urine to pass into the urethra below and then to the outside world.

During ejaculation, the bladder neck remains closed to prevent seminal fluid passing upwards into the bladder. At this time, the distal or rhabdo-sphincter opens to permit seminal fluid to pass into the urethra below the prostate prior to exiting. If the bladder neck is incompetent, when seminal fluid enters the prostatic urethra it mixes with urine already there. This abnormal situation is erroneously referred to as retrograde ejaculation, but it is really retrograde emission. In this circumstance, the body adapts and the rhabdosphincter remains contracted (with the urethra at this site occluded) and seminal fluid mixed with urine is then subsequently passed when the man next voids his urine.

The prostate has five anatomical zones, as described by McNeal et al. [26] (Fig. 2.2). The peripheral zone (**PZ**) is mostly at the back of the gland in front of the rectum (with lateral horns so that in cross-section it appears as a short "U," thickest at the bottom): it is the site where most prostate cancers arise. The ejaculatory ducts open into the central zone (**CZ**) which is an uncommon site for cancers to develop as is the transition zone which surrounds the upper part of the prostatic urethra. Those cancers that do arise in the TZ tend to have a low Gleason score/International Society of Urological Pathologists (**ISUP**) grade. Benign prostatic hyperplasia (**BPH**) (an increase in size due to an increase in the numbers of constituent cells) develops in the TZ. This noncancerous enlargement is the commonest prostatic cause of lower urinary tract symptoms (**LUTS**) in the ageing male. BPH and prostate cancer do coexist, both arising within the TZ as well as, and more commonly, due to cancer arising in other zones. The overall size of the prostate tends to increase with ageing largely due to BPH. Finally, the anterior zone (**AZ**) (behind the symphysis pubis) is another site where prostatic cancers also arise.

2.3 Normal microscopical appearances

At a microscopic level, the prostate consists of mucosal (lining) and stromal (supporting tissue) components with glandular sacs or acini lined by secretory epithelial or lining cells. Spatially, the glandular acini are arranged in tree-like formations and are lined by cells that are predominantly columnar or cuboidal in appearance with basal cells present beneath them. The basal cells are fewer in number and, together with epithelial cells, rest on a basement membrane: the basement membrane consists of type IV collagen, laminins, and other proteins to appear as a continuous lining with light microscopy [27,28]. The mucosa forms papillary projections into the lumen (or opening). These openings lead into the excretory ducts which, in turn, open into the prostatic urethra. The stromal component is larger than the epithelial component with smooth muscle cells prominent. These smooth muscle cells, which extend to merge with capsular smooth muscle cells, contract to release secretions during ejaculation. After smooth muscle, the next major component of the stroma is fibrous tissue.

2.4 Functional relationships

Prostatic fluid together with sperm form seminal fluid or semen. The prostatic fluid contribution constitutes up to 40% by volume, following contraction of prostatic smooth muscle. Approximately 70% of prostatic fluid is produced

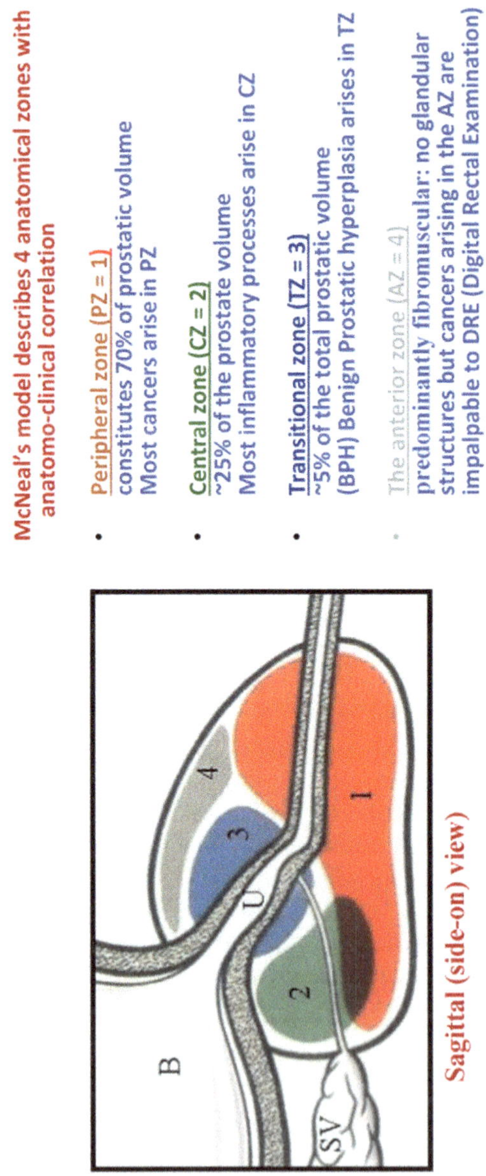

FIGURE 2.2 Diagrammatic representation of the zones of the prostate based on McNeal et al. (1988) [26].

by the PZ epithelial cells. These cells store and secrete prostatic fluid which, amongst other components in particular citrate, contain PSA, prostatic acid phosphatase (**PAcP**), and polyamines such as spermidine.

In blood and serum, prostate specific antigen (PSA) is bound primarily to the protease inhibitor α-1-antichymotrypsin and, to a lesser extent α-2-macroglobulin as well as other proteins, with a small unbound active component [29,30]. Before PSA became available for assaying, PAcP was the most used prostate cancer tumour marker but it is seldom measured now as PSA, despite its recognised limitations in detection, is a much better tumour marker. PSA, also known as KLK3, is a member of the kallikrein family of genes and proteins. PSA liquefies seminal fluid to facilitate the movement of motile spermatozoa [31,32].

As mentioned previously, the transition zone (**TZ**) of the prostate (Fig. 2.2) is the site where benign prostatic hyperplasia (**BPH**) develops. Bladder outflow obstruction due to BPH is the commonest, but certainly is not the only, cause of lower urinary tract symptoms (**LUTS**) in ageing men. This noncancerous condition can be present without causing LUTS and can coexist with cancer that mostly arises elsewhere in the prostate, particularly in the PZ: prostate cancer can also develop in the TZ but when it does, it usually has a low Gleason score/ISUP grade.

BPH causing bladder outflow obstruction is described as having two functional components viz. static and dynamic. The *static component* is largely mediated hormonally, primarily through the conversion of testosterone to dihydrotestosterone via the action of the enzyme 5-α reductase with 5-α reductase inhibitor therapy, (finasteride [Proscar]—selectively targeting isoenzyme 2 - and Dutasteride targeting both isoenzymes). These medications decrease prostatic size by inhibiting the conversion of testosterone to dihydrotestosterone (**DHT**) in prostatic epithelial cells and are effective for LUTS associated with large BPH glands with relatively high PSA levels, indicating significant epithelial cell hyperplasia. The *dynamic component* of prostatic obstruction is via smooth muscle/stroma involving α1-adrenoceptors (**α1-AdrRs**) and phosphodiesterase 5 (**PDE-5**) receptors (receptors targeted by α1-AdR blockers and members of the sildenafil (Viagra) family of drugs, respectively), amongst others.

2.5 Functional changes due to previous surgery

Many procedures have been described for the management of bladder outflow obstruction due to BPH. Transurethral resection of the prostate (**TURP**) or the "rebore" operation (Figs 2.3 and 2.4) is the best known and involves using an instrument (called a resectoscope) inserted through the urethra to resect BPH tissue from the inside so that the prostatic urethral channel is widened.

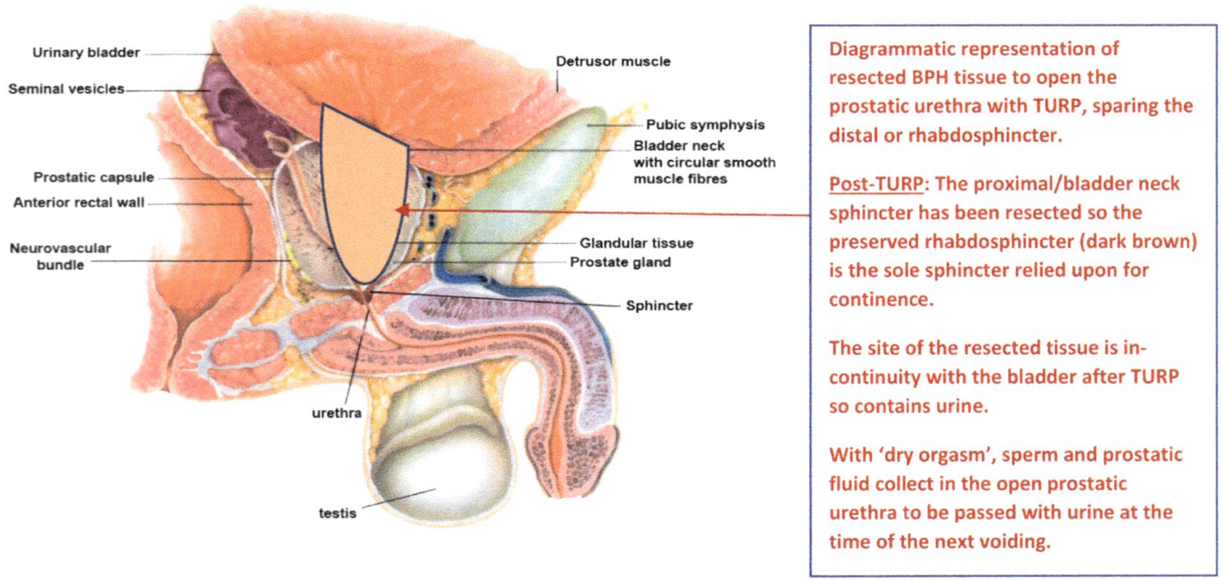

FIGURE 2.3 Diagrammatic representation of the effects of transurethral resection of the prostate.

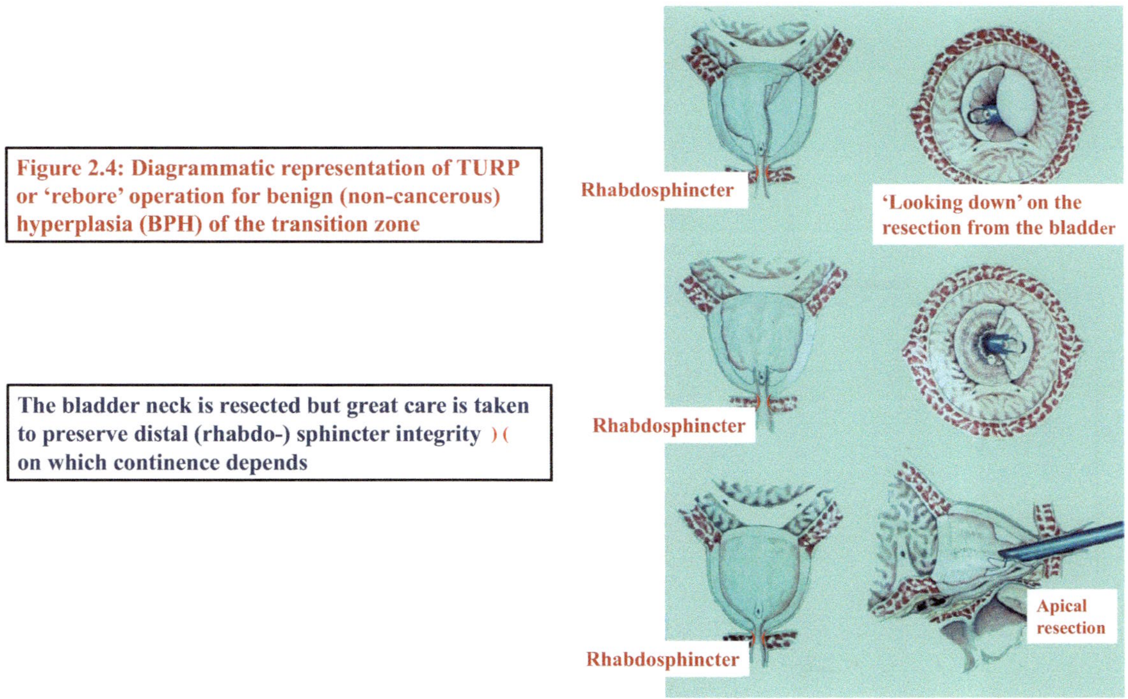

FIGURE 2.4 Transurethral resection of BPH (in the transition zone)—in progress.

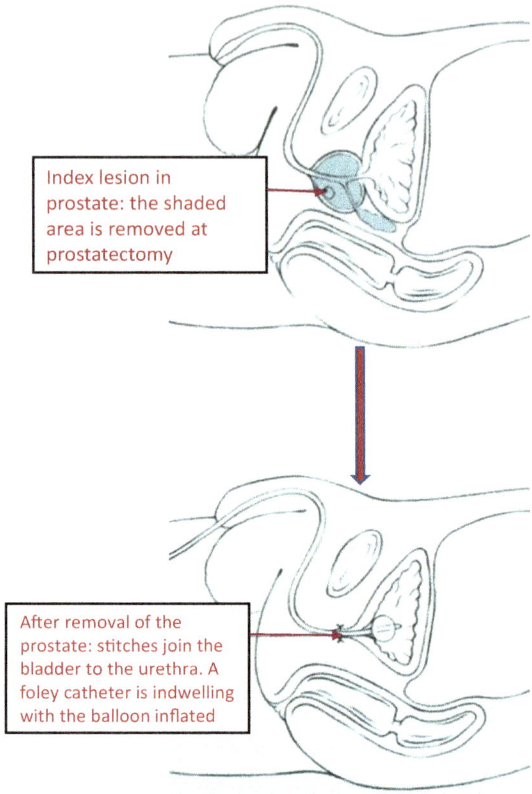

FIGURE 2.5 Radical prostatectomy.

In so doing, the bladder neck sphincter is resected but great care is directed to preserving the integrity of the distal or rhabdo-sphincter (Figs 2.3 and 2.4). This results in men remaining continent of urine but sperm and prostatic fluid mix with urine present in the open prostatic urethra so that a "dry orgasm" is experienced with sexual activity. Sperm and urine together are then voided subsequently. (During the postoperative period following TURP, epithelial [lining] cells from the prostatic ducts and bladder extend to cover the raw area exposed by resection of TZ BPH tissue).

Radical (or total) prostatectomy (**RP**) (Chapter 5) involves removal of the whole prostate for cancer with preservation of the rhabdosphincter mechanism for the maintenance of urinary continence: the bladder-neck, which is intimately associated with the upper part of the urethra, is removed along with the prostate, both seminal vesicles and the immediately adjacent part of the vas deferens (Fig. 2.1; Fig. 2.5) so that a bilateral vasectomy is also performed. Consequently, following RP, the source of prostatic fluid has been removed together with the sperm storage containers, seminal vesicles, which, along with bilateral vasectomy, ensure that all sperm produced by the testicles is blocked from exiting. Thus, with sexual activity, men experience a "dry orgasm" but for a different reason than that following TURP (as illustrated in Fig. 2.3).

Chapter 3

The background to cancer

3.1 What is cancer?

A **tumour** is a mass of abnormal tissue that forms in the body. Tumours are either benign or malignant (cancer), the difference being that benign tumours remain localised to the part of the body in which they arise. As benign tumours enlarge, they can compress and/or displace adjacent structures. Benign prostatic hyperplasia is a good example of a benign tumour that can become very large and weigh over 100 g in some instances. By contrast, malignant tumours have a propensity to invade surrounding tissues *and* to form secondary tumours in other sites in the body separated by normal tissue. In addition to being able to induce vascularisation (neovascularisation), these so-called "daughter" cancers may also demonstrate a capacity to modulate their immune environment. Secondary malignancies may become larger than their primary lesion(s) and possess an ability to grow faster. The process by which malignant tumours spread is called metastasising and will be addressed subsequently.

All cancers are the result of genes and/or gene products "gone wrong" and uncorrected by the body's checks and balances, both within cells themselves and outside them (extracellularly). As in normal development, there is an integral relationship between genes/gene products and their observable functional characteristics or **phenotype** (from Greek "pheno" meaning showing or appearance) (Fig. 3.1).

3.2 Genes and cancer

Genes consist of segments of DNA sited on chromosomes which, normally in humans number 46 (22 pairs of autosomes and two sex chromosomes which are designated X and Y for female and male genders, respectively). **DNA** (deoxyribonucleic acid) is a molecule composed of two nucleic acid chains coiled around each other in a double helix—in appearance not unlike a traditional barber's pole, based on two ascending snakes coiled around a pole. In cell nuclei, DNA is tightly coiled around histones; the DNA-protein complex is called chromatin.

DNA is the "reference book" of instructions for cells and is duplicated when cells proliferate by dividing so fidelity in DNA replication is imperative

An Insider's Perspective of Prostate Cancer: Understanding Effects, Management Options and Consequences.
DOI: https://doi.org/10.1016/B978-0-443-18709-4.00008-0

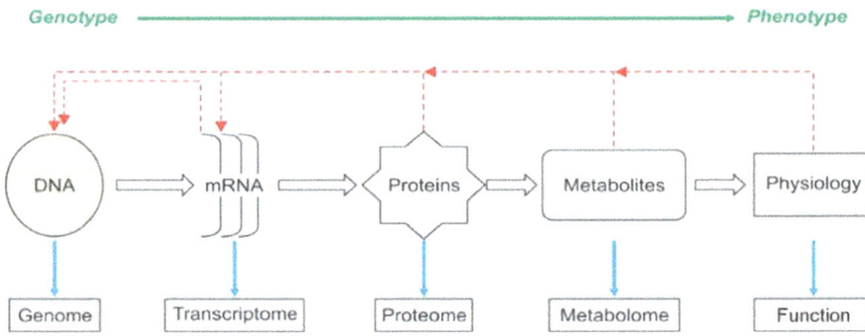

FIGURE 3.1 Flow-on effects from DNA [33]. (*Courtesy of Dr Matt Roberts*).

if cellular integrity and optimal functioning are to be maintained. Efficacy and efficiency of DNA repair genes are crucial for ensuring genetic stability with critical roles attributed to limiting cancer development and contributing to longevity in humans and other species [34,35]. Proteins are produced in cells through interpretation of the DNA code through transcription to **RNA** (ribonucleic acid), and translation to protein (Fig. 3.1). The central dogma of molecular biology depicts a relatively direct information flow across biomolecules.

RNA (ribonucleic acid) is a copy of one strand of the DNA molecule with resulting proteins the central "workhorses" in cells (Fig. 3.1). In this chain of events, RNA has been regarded as having an intermediary role. There is a variety of RNA subtypes. Messenger RNA (**mRNA**) contains genetic information in the form of **codons** (a sequence of three nucleotides that together form a unit of genetic code) transcribed from DNA in the cell nucleus. Transfer RNA (**tRNA**) transfers amino acids to ribosomes in the cytoplasm for synthesising proteins.

Another form of RNA, called **noncoding RNA**, helps in protein production by regulating gene activity. The most abundant form of noncoding RNAs is *long noncoding RNAs* which have a wide diversity of activities that include corralling proteins to parts of cells and being involved in the switching "on" and "off" of genes by recruiting specific epigenetic factors to promoter regions of genes. Noncoding RNAs interact with another subtype of RNA, **microRNAs**. microRNAs, which are also noncoding, interact with other cellular factors to indirectly affect gene expression, amongst many other functions.

In cancer, perturbations can occur with all these and other intracellular participants. For example, expression of the long noncoding RNA PCAT1 is positively linked to castrate-resistant prostate cancer (**CRPC**) progression via activation of AKT and NF-κB signalling [36] (see Chapter 7). Gene fusions are also common in prostate cancer, the most common and best-known being *TMPRSS2-ERG*: abnormal messenger RNA formed with abnormalities in transcription also occurs so that protein aberrations and functions can result with down-stream adverse consequences in terms of expression and function.

The epigenetic region or **epigenome** serves as a switch to activate or silence genes (that include epigenetic methylation and acetylation histone changes in particular), having a role in the normal development of the foetus, differentiation of one cell type as opposed to another and maintenance of cellular activity. In cancers, the well-orchestrated role of the epigenome is able to be hijacked with inappropriate reawakening of some genes and silencing of others so normal cellular functioning and appearances are disrupted. Epigenetic reprogramming in cancer genomes creates a distinct methylation landscape encompassing clustered methylation at regulatory regions modifying the suite of genes activated and silenced [37], but does not alter the integrity of the DNA transcribed.

The epigenome is susceptible to environmental and other influences, for both good and bad, in relation to cancer prevention, development, and progress. For example, in a positive way for the patient, exercise has been reported to affect the expression of inherited genes through epigenetic biochemical changes affecting the phenotype of cells [38].

3.3 Microscopic appearances

The large majority of prostate cancers are **adenocarcinomas**, arising from the epithelial (surface lining) cell component (as opposed to rare sarcomas that arise from stromal or connective tissue components), with a gradation in appearances from normal to high-grade malignancy. Features present increasingly with this gradation are increased epithelial cell numbers, variations in cell size with nuclear enlargement, and loss of the well-organised architectural pattern present in the normal prostate together with a loss of basal cells. Mitotic figures may be observed in more aggressive cancers.

Foci of atypical glands, also called "**a typical small acinar proliferation of uncertain significance,**" have features suspicious for, but not diagnostic of, cancer although they may and do accompany definite malignant changes. These encompass a variety of lesions including benign mimickers of cancer, high-grade prostate intraepithelial neoplasia (**HGPIN**), and small foci of carcinoma which, for a variety of reasons, cannot be accurately diagnosed. A much more detailed description is provided in Endotext: Prostate Cancer Detection [10]. https://www.endotext.org/section/male/.

A minority of men diagnosed with prostate cancer have a very aggressive disease that metastasises. Over time newly formed metastases may replicate much more rapidly, appear increasingly **dedifferentiated** (further away in their appearance from normal cells) and then, ultimately, change to another type of tumour appearance called **neurendocrine dedifferentiation**. These cells, many generations removed from the primary lesion whence they arose, not only appear very different to their original parent cells due to their changed genetic "make-up" but also function quite differently in the most aggressive prostate cancers by secreting neurendocrine hormones such as calcitonin, serotonin, adrenomedullin, and bombesin [39].

3.4 Cell replication and the cell cycle

The continuous process of cellular replication occurs throughout the body via the cell cycle. There are considerable differences in cycling rates in different cell types with normal prostatic epithelial cells replicating or "turning over" slowly compared with most other cells. Similarly, with few exceptions, prostate cancer cells replicate at a slower rate than those in other cancer cell types.

There are four phases described in the cell cycle with specific molecular guardians (**checkpoints**), considered essential for maintenance of DNA fidelity as transcription and other errors are not uncommon as cells cycle.

G1, the presynthetic growth phase, which occurs immediately following cell division (or Mitosis or M phase), is the longest with M phase the shortest. Invariably in the cycle, some cells die, mostly via **apoptosis** (see below), or stop dividing normally and remain arrested in G1. Cells in this quiescent or "sleeping" state, designated as being in G0, are able to re-enter G1 in the cell cycle subsequently. Some cancer cells use the G0 sanctuary to remain viable during radiation and chemotherapy then re-enter and become actively replicating in the cell cycle when the treatment stressor(s) subside(s) (Fig. 3.2).

There are many molecular checkpoints to prevent aberrant replication throughout the cell cycle with the G1-S checkpoint that monitors DNA integrity before irreversibly committing to DNA replication of particular importance. One of the best-known genes in the cell cycle is p53, dubbed the guardian of the genome. This tumour suppressor gene's normal (also called **wild-type**, as opposed to mutant) protein acts to prevent abnormalities that have developed during G1 progressing to S (DNA synthesis) phase. Once activated in response to distinct stimuli, including DNA damage or aberrant oncogene activation, wild-type (or normal) p53 induces cell cycle arrest to allow either repair (with subsequent progression to S phase) or to trigger apoptosis (programmed cell death, see below).

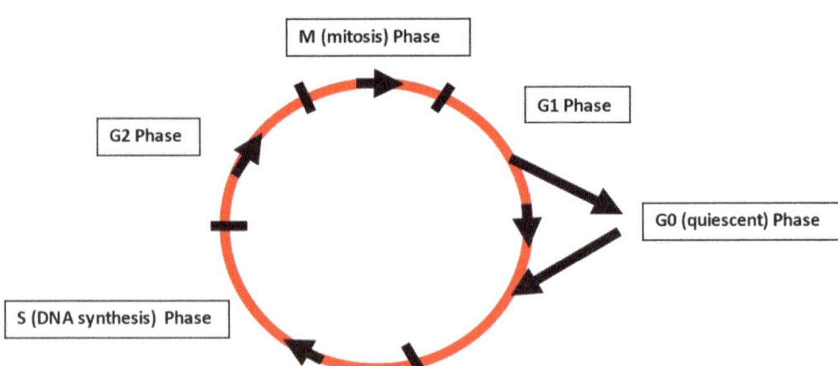

FIGURE 3.2 Cell replication and the cell cycle.

G2 (premitotic growth) phase follows S or the DNA synthesis phase, although synthesis of RNAs and proteins does occur throughout most of the cell cycle. During G2, the cell prepares for the M or mitosis phase in which cell division occurs. This is achieved by condensing chromatin (a complex of DNA and protein) resulting in long DNA molecules being compacted into denser structures. The G2/M restriction or checkpoint ensures there has been accurate genetic replication before cell division proceeds.

Cell division normally proceeds in a specific and orderly manner in M phase with spatial rearrangement of chromosomes evident in four definite steps, namely, prophase, metaphase, anaphase, and telophase. Cell division results in two daughter cells, each of which then enters G1 to start the cell cycle process again.

Throughout the cell cycle, cells are vulnerable to mutational and other changes that interfere with the evolved checks and balances that normally ensure fidelity in ensuing daughter cells. Changes to genes involved in the replication rate are a characteristic of cancer cells compared with normal cells: the more malignant a cell, the faster it can replicate.

Programmed cell death: Apoptosis (or programmed cell death), as opposed **necrosis** (cell death due to disease, injury, or failure of blood supply), occurs to both normal and malignant cells.

Characteristic visual changes with apoptosis include cell shrinkage, chromatin condensation, and formation of cytoplasmic blebs with apoptotic bodies. Apoptotic cells involute through a series of activated molecular steps starting with release of the enzyme caspase. At the end of the apoptotic process, remnants of the "imploded" apoptotic cell, contained within small membranes, become part of the "refuge collection" to be removed by immune cells such as macrophages through a process known as **phagocytosis** (ingestion and breaking down).

Autophagy, another form of cell death, occurs in cancers as a response to nutrient stress. Similar to apoptosis, autophagy is a genetically determined adaptive cellular response triggered by a range of cellular stressors such as nutrient or growth factor deprivation, **hypoxia** (oxygen lack), and reactive oxygen species amongst others [40]. Degradation of cytoplasmic organelles and other intracellular structures is the result of lysosome fusion with redundant cellular material forming in vacuoles in autophagosomes which can effectively cross-prime tumour-reactive CD8+ T cells [41] and interact with other components of the adaptive (or acquired) immune system (see section immediately below). The least common of these forms of cell death, **cell–cell cannibalism**, develops [42] in which cells of the same lineage consume their adjacent fellows.

Some notable changes with prostate cancer: Aggressive cancer cells can be extremely enterprising in turning every aspect of their environment to advantage. In addition to being able to reawaken dormant genes (for example those active in foetal development and which have become and normally stay quiescent), they can also change normal or wild-type genes to become mutant variants with dysfunctional properties.

One of the most common mutational changes in prostate cancer is loss of the tumour suppressor gene **PTEN** (phosphatase and tensin homologue), present in approximately 20% of radical prostatectomy specimens and in as many as 50% of castrate-resistant tumours. Through the cell cycle, PTEN normally prevents cells from growing and dividing too rapidly so that loss of the PTEN pathway results in this regulation being lost leading to activation of the **PI3K-Akt** (phosphoinositide 3-kinase-RAC-alpha serine/threonine-protein kinase) pathway. P13K-Akt has roles in tumour cell growth and resistance to apoptosis, with its activation in cancer cells serving to drive faster replication rates than those in their normal cell counterparts. There is therefore a strong correlation for loss of PTEN with adverse oncological outcomes so that loss of PTEN is regarded as a useful genomic indicator of aggressive disease in patients with clinically localised tumours.

Alterations to normal cell cycle machinery, such as DNA repair genes, are frequently present as cells change from replicating normally to become cancers, with previously well-controlled cellular divisional processes increasingly becoming dysregulated as cancers progress [44].

As indicated above, normal or wild-type p53 acts as a vital checkpoint at the G1/S transition so that changes affecting its integrity result in its critical function being changed: normal or wild-type p53 protein has a half-life of approximately 15–20 min compared with several hours for mutant p53, the latter having the capacity to impair if not completely abrogate the vital vetting role of wild-type p53 protein, resulting in unchecked cellular dysfunction including replication. In prostate, as in other cancers, mutant p53 proteins are present especially in the more aggressive and potentially lethal subgroups of tumours [45].

Cancers can also change replication rate by targeting chromosomes directly by a process called **chromothripsis** in which the targeted chromosomes [46] are fragmented and then reassembled in ways that promote growth. Chromothripsis, often occurs early in a tumour's evolution, resulting in many clustered structural variants arising in a single event.

As mentioned above, in addition to various gene mutations, many gene fusions have been found, the best known of which in prostate cancer is *TMPRSS2-ERG*. The androgen-sensitive TMPRSS2 is expressed in prostate epithelial cells. In prostate cancer it can fuse with ERG, a non-androgen sensitive oncogenic (cancer-promoting) transcription factor: the fusion gene is present in approximately 50% of prostate cancers and, unlike ERG alone, is androgen sensitive.

The genetic profiles of cancer cells formed as a result of various malfunctions, including failure of the normal built-in checks and balances that have evolved, are different to those of normal cells and prone to produce even more faults when they, in turn, divide. As a consequence of these many and varied disruptions, as generations of cancer cells continue to replicate their genetic profiles are at-risk of being increasingly different with implications for responsiveness or otherwise to gene and gene-product targeted therapies directed to such ever-changing and elusive moving targets.

3.5 Immunological competence

In the body's extracellular environment, the immune system has evolved to identify and destroy abnormal cells, be they cancer or otherwise but, at the same time, preserve the integrity and function of normal cells. In other words, they have to discriminate normal from abnormal including invaders from outside the body such as bacteria and viruses. The immune system is composed of innate and acquired components, the former being older in evolutionary terms. Although immunotherapies are yet to become established treatments for prostate cancer, maintaining immunological competence following other interventions is considered highly desirable for maintaining benefit from these therapies and optimising patient well-being.

The **innate immune system** is the first-line barrier and rapid-response mechanism that prevents microbial invasion. Its components include mucous secretions, neutrophils, macrophages, and dendritic cells (**DCs**) which, as well as being active in their own right, interact with B and T cells or lymphocytes (which also work together) of the **acquired immune system**. In this highly interactive and cooperative community that has evolved to maintain immunological competence, B and T lymphocytes are prominent in cancer control. (**CD** is the preceding abbreviation used universally to identify characterising cell surface molecules, including those on immune cells, which refers to classification determinant or cluster designation or cluster of differentiation.)

B Cells/lymphocytes have receptors that recognise antigens of different types. (An **antigen** is a substance that induces an immune response). Antigen-B-lymphocyte (or B-cell) binding activates the B lymphocyte which, following CD4+ **Helper T-cell** stimulation, undergoes clonal proliferation and maturation into an antibody-producing **plasma cell**. Each **plasma cell**, derived from an antigen-stimulated B cell, secretes antibodies that recognise the same antigen that was bound to the receptor of its B cell predecessor. Antibody responses can be mounted rapidly after encountering a new antigen and again to that antigen subsequently. By comparison, T-cell responses are slower to initiate and subsequently to respond.

T cells/lymphocytes are so named as they are derived from the **thymus gland**, situated in the neck and upper thorax: the thymus decreases in size, in a process known as involution, at about the time of puberty. Mature T cells are found in the blood, where they constitute 60%–70% of lymphocytes, and in T-cell zones of peripheral lymphoid organs. Most T-cells recognise only one specific antigen each via an antigen-specific T-cell receptor (**TCR**) complex, the antigen-T cell binding occurring through chemical and mechanical interactions [47]. Specialised antigen presenting cells (**APCs**), of which dendritic cells (**DCs**) are the prime example, ingest, acting a little like vacuum cleaners in this role, before processing antigens and then presenting them to T-cells via TCRs on their cell surfaces. This antigen presenting process is complex, being akin to someone having to have specially prepared food served on a platter and presented to them exclusively through designated staff before they can become active!

Antigen-specific helper T cells interact with antigen-specific B cells via an antigen "bridge": the B cell binds to one determinant (or **epitope**) on an antigen molecule while a T-helper lymphocyte at the same time recognises another determinant on the same antigen molecule. T-helper cells then bind specifically to APCs, which have taken up and processed the appropriate antigen. This process then leads to helper T cells and subsequently CD8+ve cytotoxic T lymphocytes priming [48]. Helper T-cells have a vital role in this process but can become exhausted if exposed to their target antigen(s) for too long resulting in impairment of their functioning that can lead to a reduction in CD8+ activity [49,50] which can be enduring.

Activated **cytotoxic CD8+ T cells** kill cancer cells in three ways. These are by secretion of the cytokines tumour necrosis factor-α (**TNF α**) and interferon-γ (**IFN γ**), release of cytotoxic granules from the **perforin** and **granzyme** families of proteins [also found in natural killer (**NK**) cells] and by Fas ligand (**FasL**) on the T-cell surface binding to a Fas receptor on the surface of cancer cells. (Cytokines are molecules that affect other cells: ligands are attaching molecules).

Perforin forms a pore in the membrane of the target cell allowing granzymes, proteases that cleave proteins leading to apoptosis, to destroy malignant cells. Fas ligand binding to Fas on tumour cells also induces apoptosis. Further controls on CD4+ and CD8+ cells are exhibited by Regulatory T cells (**Tregs**) which suppress activation, proliferation, and cytokine production by CD4+ and CD8+ cells. Tregs are also postulated to play a role in suppressing B cells and DCs so their modulating capacity of the immune response is quite considerable. It was thought until very recently that the role of CD4+ cells in relation to CD8+ cytotoxic T-cell killing was limited to a supporting role: however, an identifiable subset of CD4+ cells has been discovered recently that can also kill cancer cells directly [51].

NK lymphocytes are a subset of the T-cell family. NK cells constitute approximately 5%–10% of peripheral blood leukocytes patrolling the bloodstream and tissues seeking and irreversibly destroying stressed and abnormal cells, such as tumour cells. Although NK lymphocytes also perform cell killing via perforin and granzyme-mediated cytotoxicity, unlike CD8+ cytotoxic T cells they do not express TCRs so do not require antigen presentation by DCs or other APCs. In doing so NKs, as do other T cells when activated, employ aerobic glycolysis to rapidly provide glucose for energy to undertake their most important roles [52].

NK activity is regulated by a balance between signals from activating and inhibiting receptors, this balance being able to be disrupted by some cancer cells. NK cells are able to recruit other cells of the immune system, such as DCs, to the cancer microenvironment and "kick-start" a broad antitumour response by the entire immune system. Through a variety of unique receptors, which include

CD16, NK cells are able to bind to tumour-specific antibodies illustrating how the various arms of the immune system inter-relate.

In addition to helper **T-cell exhaustion** mentioned above, **immune senescence** is a gradual deterioration of the immune system with ageing characterised by shrinkage of the T-cell repertoire and a propensity to undergo apoptosis. Amongst other changes, immunosenescent T lymphocytes are less able to remove their accumulated cellular debris. This results in impaired T-cell functioning contributing to blunting of the immune response [53,54]. A subclass of NK cells known as invariant natural killer T (**INKT**) cells seek and eliminate senescent immune and other cells with irreparable DNA damage but, unfortunately, these INKT cells also become less robust with age, obesity, and chronic diseases [55]. Together, these changes can result in limitations to therapeutic strategies reliant upon and directed to reinvigorating resident T-cell populations.

Cancers utilise many ploys to prevent detection and cell killing by cells of the immune system. One example is the ability of cancer-produced vascular endothelial growth factors (**VEGF**) being able to inhibit **diapedesis** (passage of cells through the intact walls of capillaries) of T cells into tumours: selective entry for tumour cells also can occur with some tumours showing tropism for cancer cells but not immune cells. This cancer cell tropism is thought to be due to the expression of **ligands** (binding molecules) for cancer cells by endothelial (**vessel lining**) cells at preferred metastatic sites, functioning as a sort of "welcoming process." Another approach is for cancer cells to make themselves invisible to detection to immune surveillance with yet another being able to fool immune detecting cells that they, the cancer cells, are part of the body's normal cell population by masking signals and markers characteristic of cancer.

Thus, it seems that at every possible turn, the body's natural defences can be deceived and thwarted by aggressive forms of cancer making this disease such a formidable adversary, both patho-physiologically and therapeutically. However, the potential for cancers to be able to side-step the body's checks and balances have to be appreciated in the context of the natural history of prostate cancer which, for most men with this condition, means that the potential for their cancers to use these evasive actions is not realised because of interventions or because the tumour does not seem to be motivated to proliferate rapidly or spread beyond the prostate and/or to metastasise.

Chapter 4

How cancer cells progress and spread & factors in management

4.1 Feeding faster-replicating cancer cells

Replication requires energy, as does that other differentiating characteristic of cancers from benign tumours, the ability to spread and metastasise or form secondary cancers separated from the primary tumour by normal tissue. As mentioned above, cancers can subvert normal metabolic processes to serve their avaricious requirements. Indeed, metabolic alterations occur very early in the malignant process preceding evidence of histological transformation [56–58] so cells can be phenotypically (appearance with light microscopy) normal with these metabolic aberrations already in play so that both precancer and cancer cells have a growth advantage "from the word go" [59,60].

Mitochondria (singular mitochondrion) are double membraned structures in the cytoplasm of nucleated cells which, amongst other roles, serve as the energy powerhouses of cells. Energy is stored predominantly as the molecule adenosine triphosphate (**ATP**) in the matrix between the two mitochondrial membranes of these "energy packets," being produced via a series of reactions known as the tricarboxylic acid (**TCA**) or citric acid cycle, eponymously known as the Krebs cycle.

In nonprostatic cells throughout the body, citrate is an important component in the TCA cycle for ATP production as well as its roles in intracellular synthetic pathways, including lipid (fatty acids or derivatives) and cholesterol production. By contrast, normal prostate epithelial cells have evolved to sacrifice the **anabolic** (building up) role of citrate in the Krebs cycle to enable citrate, a vital component of prostatic fluid, to accumulate. Normal peripheral zone (**PZ**) cells contain a remarkable 20–70 times the level of citrate compared with other cells in the body and prostatic fluid contains citrate levels 400–1500 times that present in blood plasma [60].

In prostate cancer, this specialised adaptation for citrate accumulation is lost. Intracellular and prostatic fluid citrate levels are reduced early in prostate cancer development. Among the genes involved in these metabolic changes is the fusion gene TMPRSS2-ERG, reported to be implicated in changed citrate and spermine levels in prostate cancer cells [61]. As a result, measurement of citrate levels in

An Insider's Perspective of Prostate Cancer: Understanding Effects, Management Options and Consequences.
DOI: https://doi.org/10.1016/B978-0-443-18709-4.00002-X

prostatic fluid and seminal fluid has been proposed for early detection of this malignancy [62,63] and, as the tumour progresses, decreases in intracellular citrate and prostatic fluid citrate levels are known to parallel the change.

Cancers are able to subvert normal metabolic processes to help satiate their energy requirements. In the presence of adequate oxygen, cells will generally select the very efficient option of oxidising glucose to carbon dioxide. However, even in the presence of adequate oxygen, some malignant cells preferentially opt for the less efficient process of fermenting glucose to lactic acid, a process referred to as the Warburg Effect. By preferentially employing the Warburg effect through transcriptional reprogramming, these malignant cells give themselves a growth advantage in suboptimal conditions, common in rapidly growing cancers [64].

Mitochondria convert cellular nutrients into energy. Sugars are easiest to metabolise, more so than lipids (fatty acids and derivatives), which in turn are easier to **catabolise** (break down) then metabolise), than proteins. Recent evidence indicates that there is a cross-talk between aggressive tumours and other tissues, in particular the liver, to increase glucose production to attempt to satisfy the appetites of the most aggressive cancers [65].

Leptin is a hormone formed by fat cells with blood levels related to obesity. Leptin receptors are highly expressed on prostate cancer cells and leptin has been shown to increase cell proliferation, inhibit apoptosis, and promote **angiogenesis** (blood vessel development) [66]. Insulin and insulin growth factors, increased in prostate cancer, stimulate tumour growth via insulin growth factor receptors present on cancer cells [67]. Furthermore, insulin resistance is a common accompaniment of obesity in which peripheral tissues such as fat and skeletal muscle are less responsive to insulin, resulting in higher levels of insulin during fasting. Insulin resistance is offset by an adipokine (Greek *adipo-*, fat; *cytos-*, cell; and *-kinos*, movement) called adiponectin which, if reduced as it is in metabolic dysfunctional states including cancers, results in high levels of fasting insulin.

As a result of their voracious appetite directed to achieving a growth advantage, most aggressive prostate cancer cells not only thwart normal epithelial cell function by "out competing" those cells for available nutrients, but they also impair differentiation and function of immune cells such as macrophages [68], dendritic cells and T cells, vital to the body's attempts to curtail cancer cells' survival [69]. To compound this curbing of immune cell activities, accumulation of breakdown products of cancer cell metabolism together with hypoxia (oxygen lack), a common accompaniment of dysregulated growth, development and functions of the immune defence mechanisms are adversely affected further [70,71].

Obesity really has little going for it—except, possibly, the circumstances and pleasure enjoyed during its genesis. Recent research has demonstrated how obesity and a high-fat diet adversely affect immunological responses with some cancer cells being able to outcompete tumour-killing immune cells in obtaining fuel. When this occurs, there are fewer tumour-infiltrating CD8+T cells within the cancers themselves although the presence of CD8+T cells outside the immediate cancer environment does not appear to be affected [72]. NK cell function also is inhibited in an "obese environment" with intracellular lipid accumulation causing a metabolic paralysis leading to a loss of NK cell anti-tumour activity [73].

4.2 How cancer cells spread

Cells in the centre of cancers have less stable genomes and are more likely to spread to secondary sites. Their environment is harsher than elsewhere in a tumour due to their poorer blood supply with a resulting shortage of nutrients and hypoxia. In order to survive, subclones change to a more aggressive form and adapt so that they can survive and seek a better environment [74].

The ability to spread from their place of origin and establish satellite or secondary (metastatic) lesions at another site is a highly orchestrated process. Proteolytic enzymes secreted by tumour cells and stromal (supporting tissue) cells degrade the basement membrane and the interstitial matrix: the cancer cells change shape and become motile. This process is called epithelial to mesenchymal transition (**EMT**) [75,76].

Prostate cancer cells loosen their binding to other cells by switching adhesion molecules, in particular changing E-cadherin to N-cadherin, as well as their attachment to the basement membrane. Integrins, another class of transmembrane proteins, are also involved in the adhesion of cells to each other and to substrates. Inside cancer cells, nanovesicles have been reported to courier a specific integrin, $\alpha 5\beta 1$, from the cell surface, decreasing cellular attachment to adjacent surfaces. The intracellular nanovesicle then traffics the $\alpha 5\beta 1$ integrins to different parts of the cell surface for cellular reattachment enabling the cell to change shape and move position [77].

EMT occurs normally as the embryo develops and in wound repair. Cancer cells are able to hijack this process by activating dormant or inactive EMT genes through intermediaries (long noncoding RNAs and microRNAs). Movement proceeds by cells pushing out their membrane in the direction they intend to travel with an intracellular network of actin filaments, an important component of the cytoskeleton, following. Concomitantly, the nonadvancing part of the cell is withdrawn.

In addition to spreading directly into immediately adjacent tissues, prostate cancer cells enter the body's established drainage channels, namely the lymphatic and blood systems. Prostate cancer cells enter the vascular circulation early in the development of malignancy, but most do not proceed to become secondary cancers by escaping through capillaries (which are only one-cell thick) at other sites. This difference has consequences in cell-sampling studies as all circulating cells are not the same with genetic profiles reflecting their potentially different behavioural characteristics.

Most circulating tumour cells (**CTCs**) are solitary cells with those in clumps having a very much greater proclivity—23–50-fold—to form metastatic lesions [78]. The situation is analogous to large numbers of people in a demonstration with most protesting peacefully but groups of malevolents much more likely to move away and to cause havoc! In order to ensure that those cells with lethal potential and not the others are retrieved, techniques have been devised to select particular antigens on the CTCs and remove these exclusively for further characterisation [79].

Those relatively few CTCs destined to form secondaries proceed to bind to endothelial surfaces in small vessels and, induced by chemotactic factors originating from within the bone marrow, invade by reverse EMT (called **MET**) and other methods, with cancer cells in their elongated shapes squeezing between endothelial cells in capillaries, to move to their future secondary sites.

In anticipation, bone marrow-derived haemopoietic (blood-forming) progenitor cells home to tumour-specific premetastatic sites and form clusters, preceding the arrival of cancer cells. These "welcoming" cells provide a permissive niche for arriving tumour cells. Thus, a favourable environment for secondary cancer cells to become established is predetermined even before arrival of the tumour cells themselves [80].

Seeded cells to secondary sites may become active immediately or remain dormant in G0 (see replication and the cell cycle in Chapter 3) for a period before an undefined stimulus "reawakens" them and they re-enter the replication cycle [81–83]. All cancer cells have this ability to hibernate, a strategy employed to avoid a threat such as a course of chemotherapy, and then to resume cell cycling when the invoking threat is perceived to have disappeared.

Hypoxia, common in aggressive cancers, is a strong inducer of neovascularisation (new vessel formation). Capillaries in these new "tumour" vessels appear the same as normal capillaries with light microscopy but there are differences compared with normal capillaries at a molecular level which further facilitates the MET process [84].

4.3 Prostate cancer's preferred sites for metastases

Cancers have preferred sites to which they metastasise: this so-called **organ tropism** is thought to be through the expression of ligand receptors on endothelial cells at the metastatic sites. There are definite patterns of metastases for different cancers with secondaries occurring elsewhere than these "preferred sites" much less commonly. Prostate cancer metastasises predominantly to the bones of the vertebra, ribs, pelvis, skull, and upper femora—the so-called axial skeleton—followed by pelvic and retroperitoneal lymph nodes with fewer metastases evident in visceral structures [85].

In the lymphatic system, cancer cells travel to draining lymph nodes in the pelvis and retroperitoneum (at the back of the pelvis and abdomen alongside the

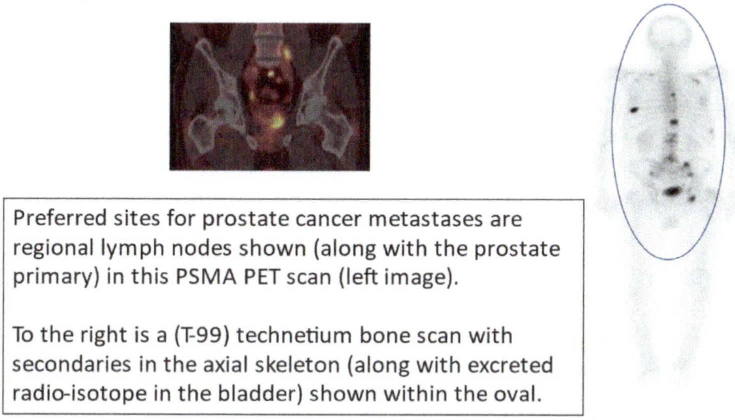

FIGURE 4.1 Prostate cancer's preferred sites of lymphatic and bone metastases.

aorta and inferior vena cava) where they are able to replace the normal residents of these nodes and are called secondary tumours or (lymph node) metastases. Vascular spread occurs in the axial or central skeleton, with secondaries forming primarily in pelvic and spine followed by upper femora, ribs, and skull (Fig. 4.1). There is a constant turnover of bone cells with a balance between osteoblasts (which produce bone substance) and osteoclasts (which resorb bone substance).

Classically, unlike the large majority of other tumours that form osteolytic skeletal metastases, prostate cancer bone secondaries almost always appear dense (**osteosclerotic**) on X-rays although osteolytic (bone deficient/absence on X-rays) metastases can less commonly occur: osteosclerotic and osteolytic secondaries can also coexist. Osteolytic bone metastases weaken the bone(s) involved predisposing to fractures and, in the spine, collapse of vertebrae. Osteosclerotic secondaries consist of woven bone (as in the embryonic skeleton). In spite of a dense radiographic appearance, osteosclerotic bone is mechanically weaker than the normal bone it replaces. This weakness is attributed to the constituents of woven bone being laid down rapidly and in a more haphazard manner.

4.4 Patient distress, psychological issues, and survivorship

Premature death continues to be significantly more common in men with prostate cancer than in the general population, particularly during the first year following diagnosis. The commonest causes are cardiovascular diseases and suicide [86]. Deaths, of course, are a crude indicator of problems belying the real prevalence and significance underlying these tragedies. Severe psychological stress is one of the main side-effects of both a diagnosis of cancer and cancer treatments, with depression affecting between 5% and 25% of cancer patients: post-traumatic stress disorder (**PTSD**) is also common.

There are many issues that arise with prostate cancer that stress patients and their loved ones, starting with the diagnosis itself but extending throughout the course of the disease [87]. A deterioration in urinary and bowel control, more common following surgery and radiotherapy, respectively [88], and erectile function after both these treatments can also contribute to adversely affecting patients' mental health [89].

Fear of recurrence is common and enduring with anxiety often palpably present when patients attend for results of PSA, biopsy, and imaging investigations [90]. Decision-related distress, uncertainty regarding treatment outcomes impacting on physical functioning and quality of life, effects on relationships as well as financial concerns affect some people much more than others.

Identifying quality of life concerns is a good start in sleuthing significant underlying concerns. However, it is one thing to recognise that psychosocial issues exist, but it is another to do something about them. Most clinics are not structured to manage these problems so, as a first step, those men who are identifiable as being at-risk should be encouraged to seek professional help directly or through patient support organisations (see **Addendum**). At a political level, advocacy for screening for distress for all patients diagnosed with prostate cancer is well and truly warranted, not just to lower distress-related mortality rates but to also improve the welfare and quality of life of all patients and their partners.

Survivorship guidelines have been established with elements identified within six domains. These are health promotion and advocacy, shared management, vigilance, personal agency, care co-ordination, and evidence-based survivorship interventions. These domains provide guidance for policymakers, clinicians, community, and consumers on what is essential for change in prostate cancer survivorship outcomes [91,92]. www.pcfa.org.au/awareness/survivorship-essentials-framework/.

Whatever the stage of prostate cancer and whichever management options are chosen, it has become clear that exercise, particularly structured exercise, is beneficial [93]. How this works remains uncertain but men with prostate cancer who exercise fare better psychologically, have less fatigue, more vitality and have a lower risk of dying from their cancer or from experiencing adverse effects from treatments [94,95].

4.5 Staging prostate cancer

Following diagnosis and after gleaning as much prognostic information as possible from the biopsy histopathology, the next step is to investigate for evidence of metastases and integrate this information with the pathology to estimate the likelihood of the cancer being localised to the prostate or whether it has spread beyond the prostate itself.

Imaging: Initially, plain radiographs (X-rays) of the axial skeleton were used but this approach lacks sensitivity as much of a bone needs to be replaced by prostate cancer for changes to be evident. CT (computerised axial tomography) is better but still lacks sensitivity with lymph node deposits needing to be $\geq$1cm before they are usually regarded as pathological with enlargement presumed to be due to prostate cancer secondaries until proven otherwise.

Technetium labelled phosphate radio-isotope bone scanning then became the standard method for detecting bone metastases. Sites of increased metabolic activity and bone turnover appear as "hot-spots" but, because other conditions such as Pagets Disease also produce increased bone turnover, CT was used to resolve this and other possible confounding causes. In addition, neither skeletal surveys with plain X rays nor Technetium-99 m bone scans demonstrate the other major site of prostate cancer metastases, lymphatic secondaries.

Prostate-specific membrane antigen (**PSMA**) is a cell-surface protein expressed by a majority of prostate cancer cells [96–98] and by tumour-induced blood vessels (but not normal vasculature) [99]. Although PSMA is also present in cells in the small intestine, kidneys, and salivary glands, its expression is at much higher levels by a magnitude of 100–1000 times in most prostate cancer cells [100].

Initial radio-isotope imaging using a radiolabelled monoclonal antibody to PSMA was used (^{111}In–capromab pendetide marketed as ProstaScint; EUSA Pharma, Langhorne, Pa): unfortunately, diagnostic results varied greatly and, overall, were disappointing, attributed to the antibody targeting a ligand-binding site on the inside of the cell membrane, a major problem when the radiotracer is not internalised [101,102].

More recently, ^{68}Ga PSMA scans have replaced technetium-99m bone scans for identifying prostate cancer metastases. As PSMA is a transmembrane protein and the ligand-binding site targeted with ^{68}Ga PSMA imaging is on the outside of cells, this approach is very much more representative than ProstaScint that mostly imaged dying cancer cells. Also, unlike Technetium bone scans, ^{68}Ga PSMA PET/CT (positron emission tomography/computerised axial tomography [also called CAT]), scans demonstrate uptake in the cancer part of the prostate itself and in most lymph node metastases as well as most bone secondaries [103–106].

Different nomograms have been available for predicting metastatic lymph node lymph node deposits for some time now. A recent international multicentre study reported that, for three of these, combining with PSMA PET imaging enhanced predictability for detecting histologically proven prostate cancer lymph node metastases [107].

Other PET imaging radiotracers have been and are being evaluated such as ^{11}C-choline with, more recently, ^{18}GF-fluciclovine, which has been reported to be of value particularly in patients with biochemically recurrent disease [108–112]. ^{68}Ga PSMA PET/CT has been shown to be cost-effective compared with other forms of imaging in prostate cancer [113] and more accurate for detecting prostate cancer metastases compared to a combination of Technetium-99m bone scans and CT scans of the abdomen and pelvis [105]. Secondaries

TABLE 4.1 Prostate cancer staging system (TNM/AJCC 2017).

Primary tumour	
Tx	**Primary tumour cannot be assessed**
T0	No evidence of primary tumour
T1	Clinically inapparent tumour not palpable not visible by imaging
T1a	Incidental tumour in < 5% of Transurethrally resected tissue
T1b	Incidental tumour in > 5% of Transurethrally resected tissue
T1c	Needle biopsy prompted by elevated PSA
T2	Organ confined
T3	Tumour extends beyond the prostatic capsule
T3a	Extracapsular, unilateral and bilateral or microscopic invasion of bladder neck
T3b	Tumour invades seminal vesicle(s)
T4	Tumour invades external sphincter, rectum, pelvic side wall
Lymph nodes	
Nx	**Regional nodes were not assessed**
N0	No regional (below level of bifurcation of common iliac arteries) nodes
N1	Regional node metastases—including pelvic, hypogastric, obturator, iliac, sacral
Distant metastases	
Mx	**Regional nodes not assessed**
M0	No metastases
M1	Metastases
M1a	Nonregional lymph nodes (outside true pelvis)
M1b	Bone(s)
M1c	Other site(s) with or without bone disease

from some highly aggressive prostate cancers, such as neuroendocrine dedifferentiated tumours, are negative for PSMA expression so that fluorodeoxyglucose positron-emission tomography (**FDG PET**) may be better for demonstrating metastases as a consequence of these tumours' very high metabolic rate.

The universally accepted staging system used to determine how far a tumour has spread is shown in Table 4.1. Staging is achieved clinically through physical examination together with findings from investigations: pathological staging is determined by the pathologist examining surgically removed specimens for analysis. The letters "c" (for clinical) and "p" (for pathological), respectively, precede the stage. For example, pT2 means that, as far as the uropathologist could determine from the submitted specimens, the tumour was confined to the prostate.

Once the stage of the cancer has been determined, the next step is to choose a management pathway.

4.6 Determining and evaluating forms of management

Tumour stratification as described above is limited as it does not take into account other factors such as PSA velocity (rate of change of PSA), the number of positive biopsy cores, percentage of Gleason pattern ≥4 cancer in the cores together with the extent of the cores involved by cancer and presence of cribriform pattern malignancy as well as molecular findings (see later). Notwithstanding the relevance of this information, the most important factors in determining the optimal management approach to be taken are the patient's wishes and his imminent commitments. Family history, opinions of partner and family, as well as personal acceptance of the potential complications of treatment and nontreatment also cannot be dismissed as it is not just a diagnosed cancer that is being managed, it is the patient whose overall welfare that is paramount.

Some treatments have been practised for such a long time that they have become regarded by default as gold standard against which other forms of management are compared. In clinically localised prostate cancer, open radical (or total) prostatectomy and external beam radiotherapy with curative intent are two such treatments.

Recognising the dilemma: There is a plethora of published data on prostate cancer treatments continually proliferating from all sorts of people and organisations, some of which is valid and helpful but some of which is not and with some quite misleading. Seemingly endless reports of studies using cell lines or with animal models in laboratory studies may or may not lead to helpful information for patients with such premature promotion by enthusiasts distracting and falsely raising hopes in desperate and distraught people. It is also of concern that a number of published preclinical and clinical reports are not able to be repeated for a variety of reasons that may or may not be justified including trial design.

It can be very difficult for an individual to critically evaluate reports of studies, let alone compare them. Researchers publish for a variety of reasons and, with differing constraints imposed, they construct their trials and studies accordingly. There are well-established criteria for ranking medical interventions with the strongest level of evidence accorded to randomised controlled trials (**RCTs**) and especially multiple RCTs. In non-RCTs for stipulated patient groups, advocates for one or other group can select the best candidates (or "pick the eyes out") of the patient population which can bias results.

In terms of evaluating clinical studies, the most important aspects to consider are clinical relevance and quality of trial design and method with a focus on practical considerations that include in particular the precision with which a study was undertaken. Many therapeutic trials, especially for advanced cancers, exclusively select participants who have a high-performance status [such as (Eastern Cooperative Oncology Group) (**ECOG**) performance status 1 or 2 (out of 6 with 6 being death)]. www.mdcalc.com/eastern-cooperative-oncology-group-ecog-performance-status. Consequently, the study populations in these trials may or may not be relevant for an individual patient, especially given often

TABLE 4.2 **NHMRC evidence hierarchy (abbreviated).**

Level	Intervention
I	Systematic review of level II studies
II	A randomised controlled trial
III-1	A pseudorandomised trial (alternating allocation or some other method)
III-2	A comparative study with concurrent controls
III-3	A comparative study without concurrent controls
IV	Case series with either post-test or pre-test/post-test outcomes

high rates of significant adverse effects which may be enduring and compound a patient's pre-existing clinical problems. In evaluating comparisons, effect size of different treatments and clinical significance are important and may be more so than statistical significance.

New therapies are investigated at different staged levels called *phases* that build on one another with each phase interrogating different questions. *Phase 1 studies* focus on monitoring safety, although safety and side effect monitoring continue in all phases. *Phase 2 studies* principally evaluate the effectiveness of an intervention and *phase 3 trials* compare the intervention with the current standard treatment. *Phase 4 studies* address long-term effects that may emerge after the therapy has been introduced into the standard therapeutic armamentarium (Table 4.2). https://guides.clarahealth.com/clinical-trial-phases/.

Also see:

CEBM: Centre for Evidence-Based Medicine
NHMRC Levels of Evidence according to Research Question. https://www.canceraustralia.gov.au/publications-and-resources/clinical-practice-guidelines
Cochrane Foundation https://www.cochrane.org/evidence
NICE (National Institute for Health and Care Excellence) https://www.nice.org.uk/guidance/published?type=apg,csg,cg,cov,mpg,ph,sg,sc
NCCN Guidelines for Patients https://www.nccn.org/guidelines/guidelines-detail?category=patients&id=49

Guidelines and journal status: In addition, specialist organisations' guidelines are a valuable reference source. They are invariably carefully considered statements based on levels of evidence [114], for example, European Association of Urology guidelines; https://uroweb.org/guideline/prostate-cancer/. In general terms, consumer statements or take-away messages included in the abstracts of manuscripts in highly regarded, peer-reviewed medical journals are couched in understandable English and carefully crafted to provide clear messages and avoid ambiguities.

Organisational guidelines are expensive and time-consuming exercises which, to be relevant, need to be up to date. Members of guideline working parties are almost always selected on the basis of their recognised expertise representing a range of fields with members' commitments mostly unrewarded financially, in the author's experience. Published recommendations are a distillation of evidence scrutinised from the scientific literature with statements made on the basis of levels of evidence, whenever these are available.

The scientific ranking of a journal, as indicated by its impact factor, is usually a fair indication of the integrity of a study published by it. These manuscripts and their conclusions are accepted only following scrutiny and approval by multiple reviewers and are valid, in the majority of instances in these high-impact publications. However, it must be acknowledged that journal assessment and selection processes are not perfect. In medicine, the top journal on the basis of impact factor is the *New England Journal of Medicine* followed by *The Lancet* with *European Urology* the premier journal in urology.

Multidisciplinary clinics: Multidisciplinary clinics are undertaken at an institutional level and, although similarly well-intentioned, are prone to subjectivity and bias depending on the specialty composition of members attending on the days of the meetings and for the cases discussed. Another limitation is that it is not possible for all aspects of a patient's care to be comprehensively considered, especially when there are time constraints for these clinics and the patient is not directly involved in the process.

As a result, it is not surprising that concerns have been expressed regarding recommendations from guidelines and multidisciplinary clinics becoming inviolable edicts [115]. This already happens to a degree with access to expensive therapies controlled by strict indications imposed by governments and organisations that regulate funding subsidisation.

In face-to-face interactions with patients, clinicians need to take into account many factors such as patients' social and personal wishes in order to fulfil the cherished role of the doctor-patient relationship with the patient treated as an individual and not a statistic. Prostate cancer and its treatments affect every part of life. Careful, sensitive, and timely communication is warranted as it can make large differences to the way men and their families are able to cope [116]. Consequently, it is not surprising that multidisciplinary clinic and other recommendations are not always followed [117].

Chapter 5

Localised treatments for prostate cancer

Following staging, the next step for tumours localised to the prostate is to stratify them in terms of their aggressiveness and their likelihood of spreading beyond the prostate subsequently as these factors are important in helping patients choose the most appropriate management approach.

5.1 Prostate cancer stratification

Prostate cancers are stratified into low, intermediate, and high-risk tumours. Stratifying criteria vary but in general, **low-risk** tumours have a Gleason/ISUP Score of $3 + 3 = 6$, a PSA $\leq$10 ng/mL and a prostate that feels benign on digital rectal examination (**DRE**). A very low-risk subgroup—**very low-risk** prostate cancer—has also been categorised in which there is a Gleason/ISUP score of $3 + 3 = 6$ in only one or two (of multiple) cores with no more than 50% of the cores involved by tumour [118,119]. **High-risk** cancers have a Gleason/ISUP Score of $4 + 3 = 7$ and everything above that (viz. $4 + 4 = 8$, $4 + 5 = 9$, $5 + 5 = 10$), or the patient has a PSA >20 ng/mL or tumour can be palpated outside the prostate on DRE. Criteria for a tumour to be considered of **intermediate-risk** is based on the above parameters falling between the criteria stipulated for low and high risk.

In addition to clinical staging/risk-classification systems including recently published STAR-CAP and the CAPRA score [122–124], commercially available gene panels are available. Notable among these are Prolaris, a gene expression panel of 31 cell cycle progression genes, reported on stored prostate tissue from biopsies and operations to independently predict risk assessment and cancer death [125–127]. The National Institute for Health and Clinical Excellence (**NICE**) advisory body reported that, by using Prolaris, treatment decisions were changed in approaching half the time [128].

Other gene panels have been reported also to have high predictive values in relation to patient outcomes from their tumours. Promark is a protein-based stratifying assay that incorporates the expression of multiple genes. Oncotype DX Genomic Prostate Score involves the expression of 17 genes to create a

An Insider's Perspective of Prostate Cancer: Understanding Effects, Management Options and Consequences.
DOI: https://doi.org/10.1016/B978-0-443-18709-4.00003-1

Genomic Prostate Score [129–131] with the Decipher genome classifier based on 22 gene expressions from prostate tumour material [132,133]. In addition to tumour tissue, increasingly "liquid biopsies" of blood and urine are being examined for genomic changes for prostate cancers [133]. Although none of these may be the last word in molecular discrimination of patients' tumours enabling better prediction and tailoring of treatment, they unequivocally indicate the potential predictive capacity of molecular characterisation in prostate cancer and herald an imminent change influencing management approaches.

Reflecting this changing clinical focus, as mentioned above, loss of PTEN is acknowledged as a key event resulting in overactive P13K signalling and the development of aggressive prostate cancer. Patient stratification based on PTEN loss or mutations in the P13K/AKT/mTOR pathway has also been advocated for selecting treatment strategies together with other molecular changes relating to tumour susceptibility to drug therapies.

5.1.1 Overview of established management approaches

Most prostate cancers are clinically localised at the time of diagnosis meaning that, as far as clinicians can tell from the pathology, clinical examination, and imaging, there is a high likelihood that the cancer has not spread beyond the prostate.

A large recently published prospective, randomised study of clinically localised disease in which 34% of participants were diagnosed with intermediate or high-risk disease (and 66% with low-risk disease) [134], concluded that more than 90% of men with localised prostate cancer are unlikely to die of prostate cancer within 10 years, regardless of whether treatment is by monitoring, surgery or radiotherapy with curative intent [135]. In other words, their results indicate that the overall likelihood of dying from prostate cancer at 10 years is the same for men receiving any of the three management approaches. A criticism of this study was the high percentage of low-risk disease patients included as, with improved diagnostic triaging now widely used, many of these men would not proceed to biopsy diagnosis today.

However, survival is only one factor for consideration. Adverse effects on sexual and bladder function are more likely to be worse for men who receive radiotherapy or prostatectomy compared with those who receive neither, but the risks of prostate cancer progression are more likely if monitoring is chosen instead of prostatectomy or radiotherapy [136]. In the ProtecT trial (cited in this and the preceding paragraph), surgery and radiotherapy were associated with lower incidences of disease progression and metastases than was active monitoring.

In addition to the well-recognised difficulties in evaluating different treatment options based on published outcomes in peer-reviewed journals for one form of management or another (as stated in the preceding section) [137], it is important to realise that meaningful published results for prostate cancer are from treatments undertaken years previously and that practice refinements in all areas continue to be made. Furthermore, various confounding factors may have been operative in the interim between treatment and follow-up assessment: among these are androgen deprivation therapies, emerging patient comorbidities, different definitions of treatment failure, administrative impacts on management and, of course, patients' levels of physical activity, healthy nutrition, and body weight.

Adding to the difficulties in comparing outcomes, a variety of study plans and criteria are used in publications to evaluate findings for different treatments. A good example of the latter is assessing disease progression post-treatment as different yardsticks are used for assessing failure of cancer control for prostatectomy and for radiotherapy (discussed below). As better modes of detecting tumour become available, it is possible to demonstrate that cancer can still be present below what were previously accepted PSA cut-offs for tumour control [138].

Thus, the crystal ball for comparing and predicting future outcomes for those recommending one treatment as opposed to another is less than clear. Furthermore, analyses of studies are from findings involving large numbers of participants, and each patient, when deciding which treatment to have, quite reasonably wants to know what will be best for *him*.

5.1.2 PSA monitoring

Following prostatectomy, any PSA detected in the blood indicates that some prostatic tissue is still present. Although very rarely, persisting benign prostatic cells or other conditions can be responsible [139–141], the presumption is that if the PSA is rising residual prostate cancer is responsible. The cause for a rising PSA could be the local recurrence of malignancy or micrometastatic disease too small to have been detected preoperatively, peroperatively, or even early in postoperative staging investigations. The PSA blood test is a poor indicator of the presence of prostate cancer pre-operatively but post-operatively, it serves as a very sensitive tumour marker.

There are some differences in studies on what constitutes biochemical failure (**BCF**) but a PSA of, or greater than, 0.2 (or even 0.25 in some studies) ng/mL is generally regarded as diagnostic [142–145]. There is a greater risk of metastases if BCF occurs within 5 years, and certainly the first 3 years: the earlier BCF occurs following surgery the greater the likelihood of metastases and death from prostate cancer [144,146,147]. Men with a BCF after 5 years or longer have a low risk of clinical progression of their disease but there remains a small risk of metastases and death.

Compared with radiotherapy, there is less room for dispute regarding the significance of the rate of rise of PSA following prostatectomy with the PSA doubling time (**PSAdt**) considered a better indicator of outcome and long-term survival than a single PSA level. A PSAdt of less than 12 months is associated with a high risk of dying from metastases which is even more likely for the small minority of patients with a PSAdt of less than 3 months [147].

Unlike prostatectomy, following which circulating PSA levels should be undetectable or negligible and remain so, radiation-induced cell death is a considerably more protracted process, so some PSA continues to circulate in the blood and may take a while (sometimes years) to stabilise. A phenomenon called "PSA bounce" is recognised which has been defined as a rise of >0.2 ng/mL. Identified following both external beam and brachytherapy, "PSA bounce" has been reported as more likely to be present in younger men, those with lower Gleason/ISUP scores, high radiation doses and is reported to be related to better survival outcomes [148,149].

The cause of "PSA bounce" remains uncertain with bacterial and radiation prostatitis [149] a late radiation effect and a delayed immune response proffered as explanations [150]. The recent widespread and routine use of adjuvant androgen deprivation therapy (**ADT**) in conjunction with radiation therapy masks the entity of PSA bounce because ADT results in a rapid reduction in circulating PSA to undetectable levels.

In addition to "PSA bounce" and effects from ADT (administered for varying periods), comparison of cancer control outcomes is also confounded by different criteria used following radiation therapy to determine BCF and outcome. The American Society of Therapeutic Radiation and Oncology (**ASTRO**) definition from 2005/6 considered a PSA rise by 2 ng/mL or more above the nadir (or lowest) PSA level to become the standard definition of BCF after external beam radiotherapy (**EBR**), with or without ADT. The 2005 Consensus Conference recommended that this definition should replace their 1996 one, further endorsed in 2017 (referred to as the Phoenix Definition) in which three consecutive rises in PSA after the nadir was deemed to be BCF [151–156]. A further confounder more recently is that the introduction of 68Ga-PSMA PET/CT technology has resulted in recurrent prostate cancer being identified at much lower PSA levels than the Phoenix definition of recurrent disease [138].

5.2 Active Surveillance

The widely accepted standard management strategy for most men with low-risk disease is Active Surveillance (**AS**), with potentially curative treatment adopted for those diagnosed with intermediate to high-risk localised disease [153–156]. It is reassuring that men with low-risk prostate cancer managed by AS who are taking 5-α reductase inhibitor therapy (dutasteride or finasteride) for lower urinary tract symptoms (**LUTS**) due to benign prostatic hyperplasia (**BPH**), are reported not to have an increased risk of developing Grade 4 or 5 disease [157].

The stratification of patients into low, intermediate, and high-risk disease is the basis for recommending management options (Table 5.1). Although criteria for inclusion in these risk categories vary, men with low-risk disease without

TABLE 5.1 An expert committee's risk classification for low, intermediate, and high-risk cancer [120,121].

Risk category	Criteria
Low	PSA < 10 ng/mL and Gleason/ISUP 3 + 3 or lower, 2 or fewer positive biopsy cores and clinical stage T1-T2a
Intermediate	PSA 10 to <20 ng/mL or Gleason/ISUP 7 (Gleason 3 + 4 or Gleason 4 + 3) or clinical stage T2b-c
(a) Favourable	Gleason/ISUP 3 + 3 (with PSA 10 to <20 ng/mL or T2b-c) or Gleason 3 + 4
(b) Nonfavourable	Gleason/ISUP 3 + 4 (with PSA 10 to <20 ng/mL or T2b-c), or Gleason 4 + 3
High	PSA > 20ng/mL or Gleason/ISUP ≥8 or clinical stage ≥T3

evidence of tumour spread beyond the prostate at diagnosis have a very small risk in the short to longer term of progression. This is even more so for a low-risk subgroup—**very low-risk prostate cancer**—that requires a Gleason score of 3 + 3 = 6 in only one or two (of multiple) cores with no more than 50% of the cores to be involved by tumour [118,119].

Of course, to call these cases cancer flies in the face of the definition which (as indicated above) states that malignant tumours have a propensity to spread from the primary place of origin to other sites in the body separated by normal tissue. However, by convention, low-risk category tumours are included under the umbrella of lesions called "prostate cancer."

5.2.1 AS and low-risk disease

Because of a very low likelihood of low-risk cancers increasing locally or spreading beyond the prostate and the fact that all treatments are associated with unwanted effects, AS is widely recommended and practised [158]. *However, it is important to appreciate that, for any man choosing AS, a significant or concerning change in clinical findings during follow-up means that the next step is for him to be offered treatment with curative intent.*

AS protocols involve monitoring patients serially with serum PSAs, prostatic magnetic resonance imaging (**MRIs**), and further biopsies (with the first within 12 months following initial diagnosis) as histopathology continues to remain the most important clinical yardstick for changing to an interventional approach [159].

5.2.2 AS and favourable intermediate-risk disease

Although contentious, many clinicians consider that these men are also good candidates for AS as a safe first-line management option [160,161], based on evidence that their 10-year mortality risk is similar to those with *low*-risk disease. Amongst many others, a UK Expert reference group in 2019 reported that a subgroup of men with Gleason 3 + 4 would be suitable for AS. Their additional requirements were an mpMRI stage of ≤T2 AND concordant biopsy and MRI findings AND PSA density (PSA divided by prostate volume) of ≤0.2 ng/mL - AND suitability for treatment with curative intent if evidence of change occurred during monitoring [158]. It is also preferred by some clinicians that the amount of Gleason pattern 4 cancer occupies <10% of overall biopsied tumour volume [162].

It is universally agreed that, practised diligently, AS is a very safe approach for low-risk cancer patients with several long-term multinational studies with large numbers of patients attesting to this contention [163–165]. However, some patients and practitioners do not comply with their AS protocol, numbers of defaulters tending to increase over time. Failure to comply with the details of the agreed AS protocol runs the risk of "missing the boat" should an adverse change occur with respect to the principle of curative intent with published results suggesting that these men are more likely to have tumour recurrence following treatment and to develop metastases [165].

At present, AS protocols remain relatively noninvasive compared with potentially curative interventions but even so are associated with adverse physical and psychological consequences. As part of the various management considerations for individuals participating in AS protocols, it is important to recognise the beneficial role that exercise can play with emerging short-term evidence that it may have beneficial effects at the level of patients' tumours [166]. It is hoped that with future noninvasive molecular profiling and continuing advances with imaging, fewer patients will be diagnosed with low-risk prostate cancer so that the numbers needing to consider AS as a management option will become greatly reduced.

5.2.3 Future AS selection and monitoring

To date, DNA [13] and /or RNA [167] molecular profiling of patients' tumour biopsies or analyses of blood or urine have not been used routinely in clinical practice in either diagnosis or monitoring. However as indicated above, it is anticipated that such changes are imminent, hopefully to be succeeded by genetic/immunological modulation of tumours noninvasively.

5.2.4 Other options for clinically localised disease

The two most common management approaches employed with curative intent for prostate cancer are radical (or total) prostatectomy and radical radiotherapy. *Although advocates for one or other may quibble about which is better, they both provide comparable survival figures at 10 years, albeit with different side-effect profiles.*

Foretelling the future accurately has always been problematic and predicting the likelihood of a cancer-curing outcome for any individual contemplating radical prostatectomy or radical radiotherapy is no different. There are multiple validated instruments that have been available for quite a few years together with more recently developed ones that include integration of evolving technological developments such as mpMRI and PSMA PET/CT [103,106,122,123].

5.3 Radical prostatectomy

Radical or total prostatectomy involves removal of the whole of the prostate along with the immediately adjacent part of the vas deferens and seminal vesicles (Fig. 2.5). The bladder and the urethra are then sutured (stitched) together, closing the gap resulting from removal of the prostatic urethra (within the prostate), thus maintaining continuity. This operation can be performed via a midline perineal incision made in front of the anus and behind the scrotum (**perineal prostatectomy**) or through the abdomen as an open operation (**open prostatectomy**), as traditional key-hole surgery (**laparoscopic prostatectomy**) or, most commonly now, as robot-assisted laparoscopic prostatectomy, commonly referred to as **robotic prostatectomy**. (This is a sophisticated form of key-hole laparoscopic surgery using enhanced instrumentation and 3-dimensional visualisation with the surgeon operating at a console separate from the operating table).

Unlike laparoscopic prostatectomy, the robotic prostatectomy surgeon has a three-dimensional view of the operating field. Perineal prostatectomy is now seldom performed in developed countries having been replaced largely by robotic prostatectomy. The benefit from removal of draining lymph nodes (lymph node dissection) at the time of prostatectomy remains contentious [168], being able to be performed via any of the three abdominal procedures. Based on a recent large, randomised study, there is no difference in the rate of biochemical recurrence comparing limited and extended pelvic lymph node dissection [169].

There has been much debate about whether robotic prostatectomy produces better results than either laparoscopic or open prostatectomy. The general consensus, based on a considerable number of published studies including one randomised controlled trial (**RCT**), is that there is no clear evidence that one is superior to any of the others [170–172] in terms of cancer control, recovery of urinary continence, or postoperative erectile function [173,174]. For best results, the most important factors are that the surgeon is experienced and that, for laparoscopic and particularly robotic prostatectomy, he or she has been trained properly and is no longer an unsupervised novice or a surgeon performing the procedure infrequently [175].

A big attraction of radical prostatectomy for some men is that complete removal of the prostate enables the whole of the prostate to be examined histologically under the microscope. As a result, a better appreciation of the cancer is provided without the (albeit now small) possibility of sampling error as can occur with diagnostic image-guided biopsies. Furthermore, if the tumour was completely contained in the gland, follow-up serum PSA levels afterwards should remain virtually zero with serial rises indicating active residual disease. Thus, following prostatectomy, many consider that they know "where they stand" in relation to the cancer.

Surgical techniques with radical prostatectomy continue to improve. Recent advances include taking into account the preoperatively measured membranous urethral length (below) and Retzius sparing surgery, which involves preserving extra-prostatic structures in the anatomical Cave of Retzius (situated behind the symphysis pubis and immediately in front of the prostate). Cave of Retzius structures includes the dorsal venous complex, preprostatic fat, and some bladder muscle. Although with limited relevance for anterior tumours, Retzius sparing techniques could potentially result in better outcomes, particularly with postoperative erectile function [176–179].

However, it should be remembered that all forms of prostatectomy constitute a significant intra-abdominal operation and are associated with the all the risks of such surgery in addition to specific risks related to prostatectomy.

5.3.1 Urinary incontinence

Urinary incontinence may be due to bladder dysfunction, impaired sphincter function or both. Reported urinary incontinence rates vary ranging from 2% to 60% with a median of 10%–15%, depending on the definition of incontinence used and its evaluation methods [173,180,181]. Generally, those men with continence difficulties preoperatively are more prone to postoperative problems depending, of course, on the cause of their urinary symptoms. Relief of preoperative bladder outflow obstruction by radical prostatectomy may actually improve symptoms and urinary control in some cases.

Lower urinary tract symptoms (**LUTS**) are more prevalent in men as they get older. LUTS may be due to the bladder muscle (detrusor) not contracting adequately, contracting prematurely of its own volition plus or minus weakness of the sphincters. As mentioned earlier, the upper or bladder neck sphincter mechanism normally remains closed, only opening when the detrusor (bladder wall muscle) contracts sufficiently strongly, voluntarily or involuntarily, to open it (Fig. 1.1). The distal or lower rhabdosphincter also normally remains closed at rest and relaxes during voiding: urine passes through the open bladder neck, then the prostatic urethra and membranous urethra before "turning the corner" to enter the anterior or lower urethra prior to exiting (Fig. 5.1).

Following prostatectomy, the rhabdosphincter is the only sphincter remaining: autopsy studies suggest degeneration and loss of these cells occurs with ageing [182,183] implying a reduced ability for the rhabdosphincter to function alone. As a result, some surgeons are reluctant to recommend radical prostatectomy for men older than 70 years of age [184]. Furthermore, the rhabdosphincter is prone to injury at operation [185] which may be inevitable for some patients

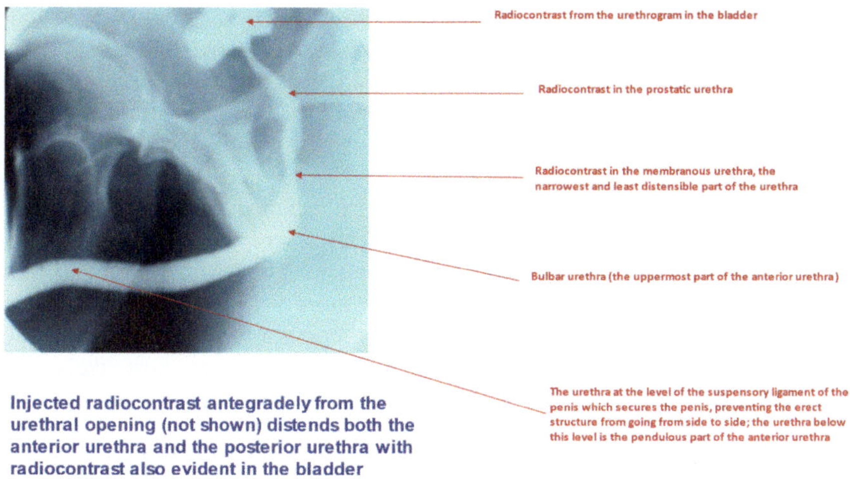

FIGURE 5.1 A normal ascending urethrogram.

as it can extend to a variable extent over the lowermost part of the prostate and the clear object of the exercise of radical or total prostatectomy is to remove *all* the prostate. Similarly, the rhabdosphincter's nerve supply can be damaged by dissection during the course of the operative procedure [186,187].

Although findings from published studies are not all in agreement, increasing age, pre-existing urinary symptoms, previous transurethral resection of the prostate (**TURP**) (see **Chapter 2).** Functional Changes due to Previous Surgery), previous irradiation to the prostate, and a high body mass index (**BMI**) indicating overweight or obesity are generally considered to potentially adversely affect postoperative urinary continence. Recently, the membranous urethral length—the length of the distance between the apex of the prostate and the bulb of the urethra (Fig. 5.1)—has been reported to relate to the likelihood of postoperative urinary continence. Membranous urethral length is able to be measured reliably using MRI preoperatively so membranous urethral length can be taken into account during preoperative counselling [179,180].

As less than perfect urinary control is common following prostatectomy, pelvic floor exercises are recommended both before and after surgery by many urologists. Although the published literature does not unequivocally endorse their benefit [188–192], most patients and patient groups are convinced that they do help considerably. Improvements in continence are stated to be particularly evident within the first 6 months after operation with little improvement likely after 1 year [188–192].

Surgical procedures for **stress urinary incontinence**—unintentional loss of urine with physical movement or activity—are directed to increasing resistance to urine flow. Injection therapy involves introducing a bulking agent into the urethral wall. Although relatively simple, infections, erosion, and extrusion are potential problems, but this approach does have the advantage of being able to be augmented subsequently. Male slings provide passive external compression to the nonpendulous urethra against the underpart of the pelvis (see Fig. 5.1): the artificial urinary sphincter involves surgical placement of an inflatable (and deflatable) cuff around the bulbar urethra with a control pump sited in the scrotum.

Barski et al. in their 2017 review [187] concluded that artificial sphincters appear to be the best option for men with severe incontinence, although associated with the highest explantation rates, and that slings seem to have acceptable efficacy for patients with mild to moderate incontinence and to be associated with lower complication rates. As most surgeons perform relatively few implant procedures annually [193], best outcomes are most likely with urologists subspecialising in this aspect of urology [187].

5.3.2 Post-operative radiotherapy

It has been estimated that between 20% and 50% of men are at high risk of experiencing progression or recurrence after their surgery [117,142,143] with any PSA detected in the blood test indicating that some prostatic tissue is still present. Although very rarely [141,194], other conditions can be responsible for detectable PSA levels in the blood [139,140] the presumption is that, especially if the PSA is rising, residual prostate cancer is responsible, this tissue not having been able to be detected preoperatively or during the operation.

An elevation in PSA following surgery almost always precedes evidence of spread or metastases by imaging and by varying periods of time, depending on the tumour's aggressiveness and replication rate in addition to how established it was in undetectable form outside the prostate preoperatively. There are differences in studies regarding what constitutes biochemical failure (**BCF**) although a PSA of, or greater than, 0.2 ng/mL [144,145] (or, in some reports, 0.25 ng/mL) [195] is regarded as diagnostic.

As mentioned above, there is a greater risk of metastases if BCF occurs within 5 years, and certainly in the first 3 years, with the earlier BCF occurs following surgery the greater the likelihood of metastases and death from prostate cancer [144,146,147]. Men with BCF occurring after 5 years or longer have a low risk of clinical progression of their disease but a small risk of metastases and death from prostate cancer cannot be discounted totally. Although it sounds obvious, it really depends on how long the patient stays alive from all causes following his surgery!

There is less dispute regarding the rate of rise of PSA following prostatectomy with the PSA doubling time (PSAdt) considered to be an even better indicator of outcome than a single PSA level. A PSAdt shorter than 12 months has been reported to be associated with a high risk of dying from prostate cancer metastases and, for the small minority with a PSAdt under 3 months there is a high risk of prostate cancer-causing the patient's demise within 6 years [147].

During the postoperative period, some men will be advised to have pelvic radiation to the prostatic bed if there are concerns that some tumour persists locally beyond what was removed at radical prostatectomy (**adjuvant radiotherapy**) or subsequently, when there is evidence of failure indicated by BCF (**salvage radiotherapy**). Indications for adjuvant radiotherapy include adverse features in the prostatectomy specimen such as positive margin(s) (tumour involving the submitted specimen's margin[s]), extra-prostatic extension (tumour seen microscopically to be extending beyond the prostatic capsule into removed periprostatic fat) or if there was seminal vesicle involvement evident.

Adjuvant radiotherapy is administered soon after surgery just after the wound has healed by contrast, the indication for the other option, salvage radiotherapy, is based on a progressive rise in serum PSA to the level decreed for a diagnosis of BCF of 0.2 ng/mL and rising [144]. Published evidence does not support an advantage for adjuvant radiotherapy compared with salvage radiotherapy for preventing prostate cancer progression [196]. By avoiding adjuvant and employing radiotherapy only when indicated by PSA changes, a number of patients can be spared the risk of attendant side effects of unnecessary radiation to the prostatic bed [197].

External beam radiation treatment can be extended to target involved lymph nodes in a form called **stereotactic radiotherapy** (**SRT**). SRT has been shown to provide a survival benefit for patients with low volume metastatic hormone-sensitive prostate cancer but not for men with high volume disease. SRT involves giving the radiation from different angles around the body with the beams meeting at the tumour so that the target site receives a much higher dose than the tissues around it [198].

There is clear evidence that combining neoadjuvant androgen deprivation therapy (**nADT**) with radiation therapy to the prostate bed lessens the likelihood of tumour progression compared with radiation therapy alone [199]. However, any form of ADT adds to the morbidity of treatment [183], as detailed in Chapter 6.

5.3.3 Sexual dysfunction

Sexual dysfunction is also common following prostatectomy, especially for those men with preoperative problems, particularly with erectile dysfunction (**ED**). Indeed, impotence has been reported to be the most common cause of diminished quality of life for patients after prostatectomy [200]. For some men, the fact that they are no longer sexually active is accepted with equanimity but for others not being able to have a firm enough erection for coitus to induce orgasm is a major issue. As indicated above, when men do orgasm following prostatectomy it is a "dry orgasm" due to the prostate and seminal vesicles having been removed and an exit pathway for any sperm permanently blocked by bilateral vasectomy (which is an integral part of the radical prostatectomy procedure) (Figs 2.1 and 2.5).

Validated instruments have been developed and are used to assess sexual function, such as the self-administered International Index of Erectile Function -5 (**IIEF-5**) and IIEF-15, expanded prostate cancer index composite (**EPIC**) scores, questionnaires designed for men who had whole gland treatments such as prostatectomy and radiotherapy. They are limited in that they focus on erectile function/dysfunction, not taking into account other aspects of sexual dysfunction such as orgasmic and ejaculatory function, sexual desire, or other masculinity/virility issues [201].

Penile erection requires a coordinated series of events involving the cavernous nerves and arteries. With sexual excitement, the cavernous nerve terminals release nitric oxide (**NO**) on to smooth muscle of the cavernosal arteries and corpora cavernosa. NO activates the enzyme guanylate cyclase which leads to smooth muscle relaxation. With smooth muscle relaxation, there is dilation of the cavernous arteries and corporal bodies, leading to enhanced blood flow and engorgement of the phallus. As the corporal bodies swell, emissary veins located along the periphery of the corporal bodies are compressed between the swelling sinusoids and the tough tunica albuginea which ensheaths the cavernous tissue. With obstruction of venous drainage, blood flowing into the corporal bodies cannot escape and a firm erection results (Fig. 5.2).

At prostatectomy, despite using a nerve-sparing technique the cavernous nerves responsible for penile erection located on both sides of the prostate may be injured. Recovery of these nerves is a slow process that can take up to 3 years [202]. During this time, there may be atrophy or shrivelling of penile tissues adding to patient dismay. Postprostatectomy external beam radiotherapy (**EBR**) may hamper ED recovery due to both a direct effect of radiation as well as ADT (chemical castration) given during this treatment [199,203].

Some prescription drugs may hamper erectile rehabilitation as they have been linked to ED. These include a number of anti-hypertensives (drugs used to treat high blood pressure), angiotensin-converting enzyme (**ACE**) inhibitors, angiotensin receptor blockers and some of the older antihyperlipidemic drugs such as fibrates [204]. Any change to drug regimens should be orchestrated by patients' medical practitioners as uninformed or uncoordinated changes may have serious health consequences. Some herbal/traditional/complementary treatments for ED may be beneficial but others may worsen ED and interact with prescription drugs, so the medical practitioners responsible for overall management should be made aware of any over-the-counter or self-administered preparations. Advertised contents of these capsules or pills are not always accurate and although they may be advertised as "natural," they may be far from safe to take or use.

Sexual rehabilitation with a doctor specialising in sexual dysfunction is often recommended. In terms of drugs prescribed to help improve penile erectile stiffness, the first-line medications are the Phosphodiesterase Type 5 Inhibitors (**PDE5I**) sildenafil (Viagra), vardenafil (Levitra) and tadalafil (Cialis). PDE5Is inhibit the breakdown of cyclic guanosine monophosphate (**cGMP**) in penile tissue to prolong smooth muscle relaxation thus facilitating erection (Fig. 5.2).

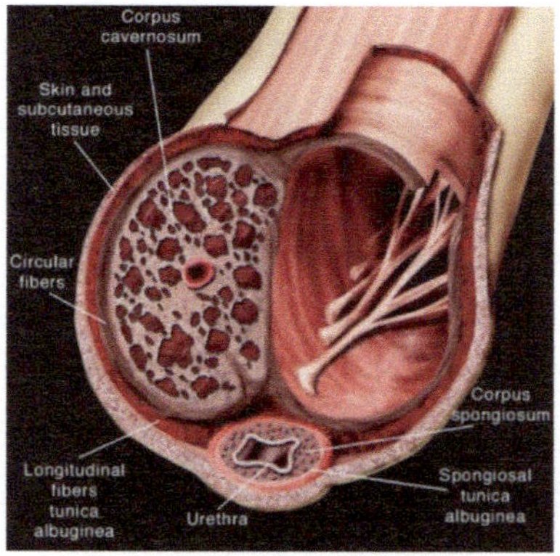

The process begins with the release of **nitric oxide (NO)** from erectile nerves onto arteries supplying blood to the penis

NO activates an enzyme called guanylate cyclase, which causes an increase in the cellular content of **cyclic GMP (cGMP)**

Arteries and erectile tissue dilate in response to cGMP, which permits increased blood flow

As blood flow increases, the spongy erectile tissue swells and compresses veins that drain blood from the penis

Compression of these veins restricts outflow of blood, leading to trapping of blood inside the erectile bodies resulting in penile rigidity

After orgasm or if sexual excitement has passed, a protein enzyme called **Phosphodiesterase Type 5 (PDE5)** breaks down cGMP, leading to reversal of this process and loss of penile erection

FIGURE 5.2 Process of penile tumescence and detumescence.

The most common adverse reactions attributed to the use of PDE5Is include headache, flushing, visual changes, muscle aches, and gastrointestinal upsets. Side effects are thought to be mediated in most cases by weak cross-reactivity of PDE5I with PDE isoforms in other tissues. If one PDE5I causes an adverse reaction, another PDE5I may be less likely to cause these symptoms. PDE5I should *not* be combined with nitrates or sublingual nitroglycerin commonly prescribed for heart pain (angina pectoris) as significant and possible life-threatening hypotension (low blood pressure) may result.

Second-line treatments include intra-penile injection with Prostaglandin E1 (**PgE1**) which leads to activation of adenyl cyclase that catalyses the transformation of ATP into cAMP which, in turn, activates protein kinases the downstream effectors leading to vasodilation. Because the release of NO is not required for these effects, any erection precipitated by PgE1 can be independent of sexual interest and desire and stimulation is NOT required for effect as it is with PDE5 inhibitors [205].

Other second-line treatments include Vacuum Tumescence Devices http://prostatecanceruk.org and, in some countries, intra-urethral instillation of alprostadil (chemically identical to PgE1: https://dailymed.nlm.nih.gov/). Surgery to insert penile prostheses is reserved for those men who do not respond to less invasive approaches [206]. All who have prosthetic devices inserted surgically should be prepared for further operative interventions at some later date if or when their devices become displaced or malfunctioning [207].

5.3.4 Psychological issues with prostatectomy

The impact of prostatectomy on men's psychological state cannot and should not be overlooked. The knowledge that their surgery has removed their ability to spontaneously procreate, whether they intend to do so or not, may make some feel that they are a "less of a man" as a result of having had the operation. Performance anxiety and ED, no matter how understanding a partner is, may also affect self-esteem and how many perceive themselves. These feelings may be exacerbated by penile shortening, a loss of libido and painful or less satisfying orgasms following surgery. The role of partners is a critical but an often-overlooked aspect influencing postprostatectomy patients' overall recovery.

The ability to return to work, socialise, and to function physically at a level acceptable to each individual is also important. Severe ongoing incontinence hampers confidence in these aims especially if there was an expectation preoperatively to the contrary. The fear of becoming a "smelly old—or not so old"—man resulting from even slight urinary or bowel dysfunction can weigh heavily on patients' minds. In the only randomised controlled study of robotic and open radical prostatectomy, one in five reported increased psychological distress at 24 months, highlighting the need for regular distress screening beyond the early treatment phase [170,171].

As mentioned in Chapter 1, many men undergoing radical prostatectomy have significant comorbidities preoperatively. In a recent overview of cognitive decline associated with anaesthesia and surgery in older patients, the authors speculated that, in addition to well-established comorbidities such as hypertension, poor nutritional status, and suboptimal diabetic control, neuroinflammation due to perioperative stress, underlying vascular disorders, and undiagnosed neurodegenerative conditions were also potential contributors [208,209].

In their report from the American Society of Anaesthesiologists Brain Health Initiative Summit in 2018, Mahanna-Gabrielli et al. reported that up to an estimated 65% of patients 65 years and older who underwent noncardiac surgery experienced delirium with long-term cognitive decline reported for 10% [209]. To date, there would appear not to be evidence to suggest that any specific anaesthetic agent should be indicated or avoided for preserving cognitive status [208].

5.4 Radiation therapy

Radiation therapy for cancer (also known as radiotherapy) commenced at the beginning of the 20th century following pioneering research by Nobel Laureates Madame Marie Curie and her husband Pierre. Radiotherapy involves the use of high-energy particles or waves from the upper end of the electromagnetic spectrum, such as X-rays, gamma rays, electron beams, or protons, to destroy cancer cells.

Radiation of these various types appears to be equally effective in cell killing capacity although proton therapy, in particular, has been promoted as better than radiotherapy from other sources. Charged proton particles stop at a fixed distance as they pass through tissues, this property highlighted as an advantage in comparison with the gradual fall-off of energy with photons from X-rays. A recent review, however, found that outcomes to date with proton therapy in prostate cancer have been similar in terms of efficacy and toxicity compared with other forms of radiation therapy but at a greater financial cost. However, as always, "the proof of the pudding is in the eating" so outcomes from RCTs are keenly awaited [210].

It is during the G2/M phase of the cell cycle that cells are most radiosensitive, less sensitive in the G1 and least sensitive during the latter part of the S phase of the cell cycle [211] (See **Chapter 3**, Replication and the Cell Cycle). In addition to a direct effect on DNA, there is an indirect effect via reactive oxygen species (**ROS**) that form as a result of radiolysis of water and oxygen to also damage cellular DNA leading to cell death. However, not all prostate cancers respond equally to radiation therapy with molecular changes such as mutations in p53-inducing resistance.

Ionising radiation doses are measured in **grays** (abbreviation **gys**) with one gray defined as the absorption of one joule of radiation energy per kilogram of matter [212].

5.4.1 Adaptations

Because prostate cancers are relatively radio-resistant, to achieve effective tumour control a high dose of radiation is needed within a limited range to optimise tumour-killing but to avoid damage to normal tissue. In practice, the amount of radiation delivered varies in doses per session, number and frequency of sessions, and cumulative totals given [197,213,214]. Its delivery needs to be modified if the patient has had a previous TURP, or if radiation is delivered to the tumour bed for suspected residual pelvic (extra-prostatic) tumour following prostatectomy. Similarly, the total dose of external radiation administered to augment whole gland (local) brachytherapy is modified.

5.4.2 Sensitisers and hypoxia

Many candidate radiosensitisers including a number of chemotherapy drugs have been used in prostate cancer in the past with probably the best known being misonidazole. Misonidazole and metronidazole (Flagyl) are members of the nitroimidazoles family of drugs, active against bacterial anaerobes and protozoa, that are also hypoxic cell radiosensitisers [215]. After initial promise in conjunction with radiotherapy in prostate cancer, misonidazole fell out of favour as the clinical benefit was relatively small and its use was associated with troublesome peripheral neuropathy in some patients which could be progressive if the medication was continued [216]; however, there is a renewed interest in other members of this family currently [216].

Hypoxia (deficient oxygenation) is common in rapidly growing tumours and induces radio-resistance [217]. Several factors are considered to contribute to the development of tumour hypoxia which is associated with the presence of aggressive cancer cells [218]. A variety of efforts has been expended in trying to overcome this problem, a prominent focus recently having been exploiting the property of the hypoxic drug pimonidazole as a marker of tumour aggressiveness [218]. Pimonidazole accumulates in hypoxic tissue [219], binding to thiol-containing proteins, reducing their radioprotective properties and thus sensitising cells to radiotherapy.

The standard method is to tackle the issue indirectly by giving patients androgen deprivation (**ADT**) periprocedurally to shrink the prostate cancer and thus hopefully reduce the likelihood of hypoxia. Higher doses of external beam radiation therapy (**EBRT**) can be targeted to the location of the highest Gleason (**Index Lesion**) using a technique called intensity-modulated radiotherapy (**IMRT**). A further innovation is to commence a programme of physical exercise immediately prior to commencement of radiation therapy, based on findings from animal and human studies that have demonstrated associated increased blood perfusion accompanied by a significant reduction in hypoxia within the prostatic tumours themselves [220,221].

TABLE 5.2 Type of radiotherapy in relation to clinical risk stratification.

Treatment options for localised prostate cancer: stratified by risk group	Treatment options
Low-risk—if procedural treatment is warranted	**If expected patient survival ≥ 10 years**, active surveillance, external beam radiation or brachytherapy
Favourable intermediate-risk	**If expected patient survival ≥ 10 years**, active surveillance, external beam radiation or brachytherapy
Unfavourable intermediate-risk	**If expected patient survival ≥ 10 years**, external beam radiation with ADT, external beam radiation and brachytherapy boost and ADT
	If expected survival < 10 years, observation, external beam radiation with ADT, and external beam radiotherapy with brachytherapy boost and ADT
	If expected survival > 5 years or symptomatic, external beam radiation with ADT, external beam radiation and brachytherapy boost and ADT
High-risk	**If expected survival ≤ 5 years and asymptomatic**, observation, external beam radiation and ADT

There are relatively few lymphocytes within prostate cancers and those that are present are found mostly at tumour margins. Although radiation is a powerful immunosuppressant for nontargeted tissues with circulating lymphocytes particularly sensitive to radiation [222], a recent comprehensive analysis of prostatic cancers before and after high dose radiotherapy reported lymphocyte, macrophage, and antigen presenting dendritic cell density numbers and their activation status within prostate tumours to be *enhanced* [223].

It is becoming increasingly clear that a highly complex interplay exists between immunological cells, molecular changes within tumours, and drug effects in patients receiving radiation therapy [224]. Consequently, there is a recognised imperative to understand these factors better so that other treatments can be integrated in an informed and considered manner with radiation schedules in individualising precisely delivered radiation regimens [225].

5.4.3 Which type of radiotherapy?

After having decided to have radiotherapy, the type of radiotherapy needs to be selected. This is often advocated on the basis of clinical tumour stratification such as proscribed but, of course, tumour stratification is *not* the only criterion. The recommendations in (Table 5.2) have been modified from a recent review [226].

5.4.4 Androgen deprivation therapy (ADT) with radiotherapy

The sequencing and duration of ADT in relation to radiotherapy remain contentious issues [214,227–229]. Neoadjuvant ADT (that is ADT commenced before radiation treatment begins) is regarded by many as routine for all forms of radiotherapy administered for prostate cancer with continuation of ADT for 18 months or 36 months more effective in delaying clinical evidence of disease progression and benefitting survival than a shorter period of 6 months administration [228,229]. However, although ADT does enhance the tumouricidal effect of radiotherapy, it does result in additional unwanted effects adversely affecting the quality of life [230,231]: these are addressed in more detail subsequently (in **Chapter 6**).

Unlike other toxicities associated with prostate cancer treatment, the side effects attributable to ADT when given to enhance the effects of radiotherapy, are usually regarded as transient and are generally considered likely to resolve after discontinuation of ADT and with a return to "normal" testosterone levels. However, testosterone recovery rates do vary. Older men, those with low pretreatment testosterone levels and men who receive ADT for longer periods of time are more likely to have slower recovery rates. One study reported that, 24 months after stopping ADT, approximately one-quarter of men had failed to return to their pre-ADT testosterone levels with one-tenth continuing to have castrate testosterone levels [232].

In addition to combining ADT with radiotherapy, other combination treatments are being examined. These include (amongst others) docetaxel with indications that the addition of this drug may provide a long-term benefit in terms of tumour control for those patients with unfavourable risk disease and a PSA of less than 4 ng/mL [233].

5.4.5 External beam radiotherapy

- **Fractionation and hypofractionation**

The commonest form of radiotherapy is delivered from outside the body known as **EBRT/EBR** or teletherapy. Cancer cells replicate faster than normal cells so are more vulnerable to radiation-induced cell death than their normal counterparts although, compared with most cancers, prostate cancer cells replicate slowly. Treatments are fractionated and given over many weeks in short bursts so that as many malignant cells as possible are exposed when they are most vulnerable, ideally when in G2/M of their cell cycles. Less malignant cancers are slower to proceed through their cell cycles so careful planning of each individual's radiotherapy programme is required.

EBRT doses of 2 Gy per fraction per day are used conventionally with total doses >70 Gy required to be effective [223]. Hypofractionation, made possible by recent improvements in tumour localisation and dose delivery, involves larger radiation doses per treatment because of prostate cancer's greater susceptibility to larger fraction sizes [234]. An added benefit for patients is that hypofractionation translates into fewer radiation unit visits for treatment.

- **Conformal radiotherapy**

With EBRT, the beam is usually directed to the whole of the prostate with a margin of 5–10 mm [235]. A form of EBRT, conformal radiation therapy, uses images from computerised tomography (CT), MRI, and positron emission tomography (PET) scans to improve precision in planning the treatment area. Images obtained are then analysed through a computer programme to conform to the shape of the tumour and the prostate.

- **Maintaining accurate targeting and gating**

Considerable effort is expended to avoid damaging structures immediately adjacent to the prostate, in particular the rectum, bladder, and urethra. This is achieved by using several windows to target the gland to minimise scattering effects to normal tissues from intervening and adjacent structures [235] with repeated or continuous imaging employed to ensure that the radiation beam remains on-target. To overcome the fact that the prostate moves slightly with positional internal changes of anatomy, radio-opaque fiducial gold markers are often injected into the gland in three separate noncoplanar sites so X-rays and other forms of imaging can be used repeatedly during treatment for checking the prostate's position for radiation delivery as well as for volumetric targeting.

As mentioned above, the prostate is a mobile structure within the pelvis with potential for a directed radiation beam to become "off-target" during a treatment session with one study reporting movement of up to 2.2 cm. Computer programmes that prevent radiation being delivered when the prostate is "off target," called beam **gating** may sometimes be used with a view to minimising damage to nonprostatic structures [236,237].

- **Unwanted radiation effects**

The total dose of radiation given with EBRT alone tends to be limited by the risk of damage to surrounding structures. Radiation induces an endarteritis (inflammation of the lining of arteries) causing reduced blood flow and results in fibrosis as an unwanted long-term consequence. As a result, there is an attendant loss of tissue suppleness and reduced wound healing should surgery to the area be undertaken subsequently.

The anal canal and rectum are particularly vulnerable to unwanted radiation damage with proctitis a well-recognised postradiation problem. Bowel toxicity, which can become totally incapacitating for a very small proportion of patients, occurs in two phases, acutely within 3 months and late, after 3 months following treatment [238]. It manifests itself with mucous overproduction, increased bowel movements and, in a small percentage of patients, bleeding from abnormal radiation-induced blood vessels. Pre-existing haemorrhoids also can become more troublesome at this time. Acute radiation toxicity is reported to predict the likelihood of the development of late radiation toxicity [238].

The abnormal vessels that form following radiation are known as radiation telangiectatic vessels. They also are found in the bladder as well as the rectum and can bleed continually seemingly without any apparent triggering factor. They are notoriously difficult to manage and can ooze to the extent that blood loss becomes life-threatening such that exenterative surgery may be required.

As with anaesthesia and surgery, a proportion of patients receiving radiation with ADT has been found to experience adverse cognitive changes, both objective and subjective [239]. Associated changes include increased levels of anxiety/depression, fatigue, and disturbed sleep. Inevitably affected patients experience a worse quality of their lives as no doubt do their loved ones. Several studies have shown that ADT, independent of the effects of radiotherapy, contributes to worse sexual function with the combination also resulting in a worse quality of life [230,231]. In addition to a variety of adverse effects, in terms of sexual dysfunction men could be considered to experience a "triple whammy" as a result of ADT in that they have a decreased ability to reach orgasm, decreased quality of erections and decreased sexual function.

- **Spacers**

To minimise rectal damage from radiation, physical means of displacing the rectum away from the radiation beam have been developed with variable clinical uptake. Rectal balloons, introduced into the rectal lumen during each radiation treatment to try to keep the back wall of the rectum as far away as possible from the prostate to be irradiated, have limited patient compliance and do nothing for the anterior or front wall of the rectum.

More recently, perirectal spacers, consisting of resorbable materials, have been injected through the perineum to develop the potential space between the front wall of the rectum and the back wall of the prostate. This results in *both* front and back walls of the rectum being displaced backwards from the prostate. This developed space contains (and retains) injected material which is slowly resorbed by the patient's body over the following 12 months and following completion of the course of radiotherapy. Perirectal spacers are injected using imaging from transrectal ultrasonography (Fig. 5.3) with the procedure able to be performed at the same time as when fiducial markers are introduced. Although the concept is sound, insertion is an operative intervention and not without risks, with remaining uncertainty as to whether injected spacers will prove to provide measurable benefits therapeutically and financially.

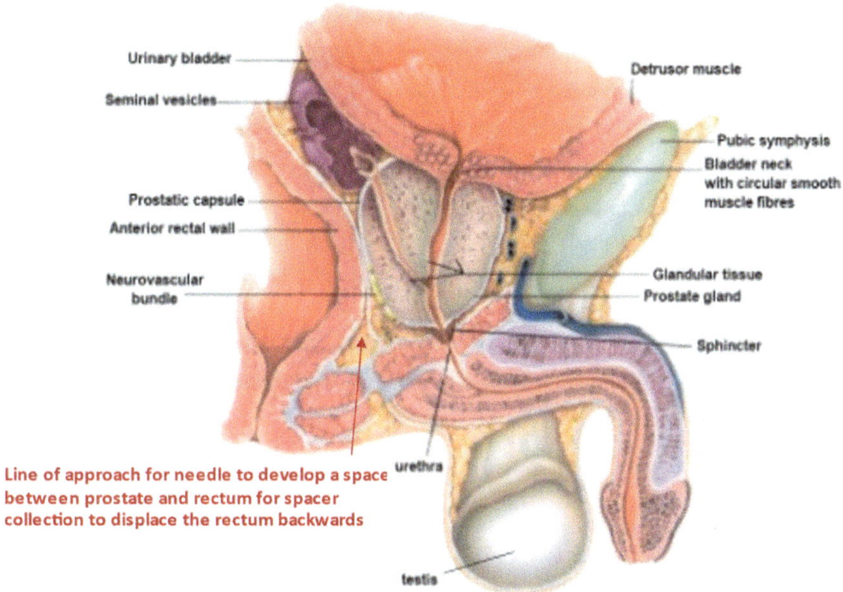

FIGURE 5.3 Direction of injection of hydrogel material as a S pacer for radiotherapy.

- **Radiation versus Surgery**

Taking into account patient selection, with many men who receive radiation therapy older and often less fit, studies indicate that *radiation therapy is associated with worsened bowel symptoms and prostatectomy is more likely to adversely affect urinary control and erectile function*. Both treatments are reported to be equally effective in terms of cancer control [240] with comparable adverse psychological effects with respect to cancer-related quality of life, depression, and anxiety in particular [89,226].

5.4.6 Brachytherapy

Brachytherapy or implant radiotherapy is a localised treatment for prostate cancer. There are two main types described, low-dose rate (**LDRB**) and high-dose rate (**HDRB**). With LDBR, radioactive seeds are placed into the prostate with needle insertion via a grid template-directed by transrectal ultrasound imaging of the prostate during the procedure. The seeds are retained permanently, although they remain radioactive for only a short time (Fig. 5.4). HDRB involves temporary radiation delivery, with no retained radiation sources following the procedure.

- **Low dose rate brachytherapy (LDRB)**

Low dose rate brachytherapy has a well-established role and may be used for patients with low-risk prostate cancer for whom AS is not considered appropriate

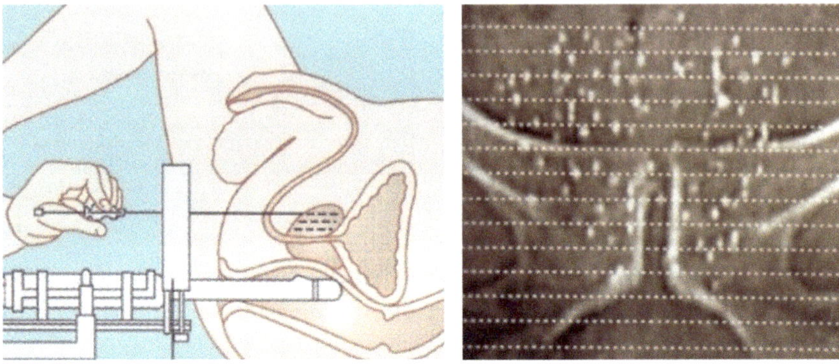

FIGURE 5.4 Low-dose rate brachytherapy with implanted radioactive seeds in the prostate. Radioactive seeds being implanted over grid with an ultrasound probe in the rectum to provide image guidance: a subsequent pelvic X-ray shows the distribution of radioactive seeds.

and for men with favourable intermediate-risk prostate cancer who decide to have this active intervention. An advantage of LDRB, which provides a high dose of radiation over an extended time—typically 145Gy over weeks to months—is that it is associated with a lesser likelihood of radiation damage to immediately adjacent structures, namely rectum, bladder, and urethra compared with other forms of radiotherapy [223]. A further radiation boost by external beam radiation therapy is sometimes focussed on an index lesion adjunctively or given to the whole prostate.

Outcomes for LDRB are comparable with those from EBR and radical prostatectomy for low and intermediate-risk prostate cancers, but with a lower risk of urinary incontinence [241–243]. Although LDRB boosted with EBR (for higher risk disease) has been shown to decrease the chance of cancer recurrence, the combination is associated with a higher likelihood of post-irradiation complications. These include an increased risk of temporary urinary catheterisation with incontinence pads being required [244], clinically meaningful-worse urinary irritative and bowel function for over 3 years and bothersome urinary symptoms enduring at 5 years [245].

Frequency and urgency are not uncommon symptoms experienced soon after brachytherapy treatment and may last weeks to months but tend to improve with time; however, incontinence worsens over the longer term with over 10% of patients in one study requiring the need to use an incontinence pad after 10 years or so with another reporting that this rose to ~40% for those men who had a TURP previously [246,247]. Difficulty with erectile function assessment is compounded by background noise from increasing ED associated with ageing and potential confounding effects of ADT when prescribed: however, the general consensus is that erections are less likely to be adversely affected by LDBR than they are by prostatectomy [244].

It must be remembered that low dose rate brachytherapy actually delivers a very high total dose of radiation, and the dose is not innocuous nor, as indicated above, the risk of unwanted effects absent.

- **High dose rate brachytherapy (HDRB)**

High dose rate brachytherapy is generally restricted to patients with intermediate-risk or high-risk prostate cancer with the intention of cure. HDRB delivers high doses of radiation over a very short time, via implanting needles introduced into the prostate under anaesthesia. The prostate is mapped by USS or CT, and a treatment plan to formulate a highly conformal radiation dose in 3D developed with a high-energy radiation source then driven down each needle under computer control to typically deliver >10Gy of radiation over 10–20 min. These large fractional doses are very effective at overcoming the resistance of prostate cancers to small doses per fraction [223] so even while a low total radiation dose is delivered by HDRB, it equates to a much higher dose in terms of cancer control.

The largest experience for HDRB is in combination with EBRT as a boost for high-risk cancer, with one or two implants introduced typically 2 weeks apart. Neoadjuvant ADT is usually commenced some months prior, as is standard with EBRT. There is also an expanding experience for using HDRB without EBRT via two implants for intermediate-risk prostate cancer. This is obviously convenient for the patient and avoids adjunctive ADT in most instances.

Outcomes are similar to those from surgery in terms of freedom from recurrent disease. The same advantages as for LDBR applies for HDRB, in terms of benefits for sparing surrounding structures and side-effects. Because there is less trauma than with LDRB, fewer patients are likely to require alpha-blocker therapy to lessen bladder outflow resistance for difficulty with bladder emptying postprocedurally. LDRB treatment may involve an overnight stay in hospital [248].

5.5 Focal therapy as a primary intervention

Improved imaging of the prostate with MRI (and CT-PSMA) now permits prostate cancers not only to be identified precisely but also characterised in terms of their degree of malignancy prior to confirmatory precision biopsies of these so-called index lesions. As a result, an identified small, localised high-grade tumour may be able to be managed with confidence by treatments directed specifically to this focus, sparing the rest of the prostate and thus lessening the likelihood of unwanted effects [249].

Focal therapy is not a new concept. Despite being an attractive option in theory because of its safety and limited invasiveness, there is a paucity of long-term published evidence from trials; those studies that have been published have varied considerably in their protocols and patient follow-up. Furthermore, there is a lack of consensus of how to monitor patients after focal therapy (PSA or imaging-based review), what constitutes recurrence with various studies using the Stuttgart definition (a rise of $\geq$1.2 ng/mL above the nadir PSA, the Phoenix definition (a rise $\geq$2ng/mL above the nadir PSA), and the Horwitz definition (2 consecutive increases of at least 0.5 ng/mL, backdated) and what would be the appropriate timing and type of further treatment. Consequently, focal therapies have not yet received widespread endorsement as first-line treatment for prostate cancer expected by regulatory authorities and guidelines [250].

The many forms of focal therapy include focussed external beam radiotherapy (EBRT) with the so-called "**Cyberknife**" system, localised brachytherapy, cryoablation, photodynamic therapy, high intensity focussed ultrasound (**HIFU**), and **electroporation** with what has been dubbed the **Nanoknife**.

Cryoablation causes prostate tissue death by freezing; photodynamic therapy involves passing laser energy through carefully placed optical needles in the prostate following intravenous injection of a photodynamic agent to produce destructive reactive oxygen species (**ROS**) within cells [251]; electroporation employs pulses of electric current to destroy cells via cell membrane damage and HIFU heats tissue precisely resulting in a form of cell death called tissue necrosis (from the Greek "dead body"). Necrotic tissues autolyse and, following resorption, result in fibrosis (scar tissue) formation.

A recent review and meta-analysis of HIFU concluded that it can be regarded as superior to prostatectomy in terms of urinary and sexual outcomes but there remains clinician uncertainty regarding long-term oncological outcomes and effective management of local recurrence. Other unresolved issues include whether adjunctive ADT is beneficial for cancer control for high-risk disease treated with HIFU and whether pretreatment TURP might reduce the likelihood of postoperative acute urinary retention [252].

The National Institute for Health and Clinical Excellence (**NICE**) has issued guidance for the use of HIFU in the NHS in England, Wales, Scotland, and Northern Ireland for localised prostate cancer. The NICE statement indicates that there are no major safety concerns with HIFU including with re-do treatments [253,254]. However, the statement continues to caution that, as prostate cancer is often multifocal, the procedure should only be used with special governance arrangements, consent, and audit or research.

Certainly, because focal treatments are limited to part of the prostate, thorough preprocedural assessment is essential with ^{68}Ga-PSMA PET/CT reported to be more likely than mpMRI to demonstrate multifocal disease [103]. Ongoing close follow-up cancer monitoring of all men undergoing focal therapy is particularly important as further clinically significant tumours may emerge over time.

Electroporation has not been used as long as HIFU for prostate cancer focal therapy. However, evidence to date indicates that men experience little, if any, deterioration in their genitourinary function as a result of this form of treatment too [201,254]. Published reports are too short to comment on its efficacy in terms of cancer control but in-field clearance of cancer in the treatment zone at the 12-month post-electroporation biopsy has been reported to be very high [255]. As with HIFU repeated (re-do) treatments are able to be administered as required so that, in theory, a further tumour arising the prostate can be addressed as it arises with the same applying to index lesions subsequently identified as having been incompletely treated initially [256].

5.6 Fear of recurrence

Being told that residual cancer persists and/or secondaries are present is received invariably as devastating news, particularly if unwanted effects from a local treatment persist. However, it is necessary to approach the problem in the same reasoned and objective manner as was exercised in determining the initial localised treatment. One ploy advocated involves likening life and its problems to a game of cards. It is not so much the cards you receive but how you play them that makes the difference.

The expectation with radical prostatectomy is that the whole of the prostate has been removed, although in some cases this may not have been possible due to preoperative inflammation or other factors influencing dissection. As a result, the natural presumption is that a detectable PSA indicates residual disease and that it reflects the highest Gleason score reported in the submitted prostate.

Studies have shown that the pattern of treatment failure is predominantly local with a surprisingly low incidence of metastatic failure, even in high-risk disease. Postoperative radiotherapy to the prostate bed, almost always combined with a course of ADT, is very effective treatment with approximately a 50% reduction in postprostatectomy BCF [257].

By contrast, radiotherapy administered with curative intent coupled with ADT results in PSA levels plummeting to negligible levels. Subsequently, with increased testosterone production following cessation of ADT, serum PSA levels rise before stabilising at a new (all be it at a lower) level than initially due to remaining viable prostatic tissue. (The effects of ADT are addressed in detail below in **Chapter 6**).

Relevant questions include:

- Is a persisting PSA due to benign prostatic tissue or, more probably, prostate cancer and if the latter, does it reflect the highest Gleason score in the primary specimen?

PSA in blood has a half-life of 2–3 days [258] so a test taken too soon following surgery will record a falsely high baseline post-surgical level with the nadir (or lowest level) taking longer to be reached if the preoperative PSA level was high. Of more relevance is the rate at which the PSA rises or the PSA doubling time.

- As an unlikely though possible scenario, could a continuing rise in PSA be influenced by pelvic instrumentation or trauma as these may affect any residual prostatic tissue, benign or malignant or due to another cancer?
- Is there any evidence of secondaries with imaging, the likelihood increasing as the PSA continues to rise?

Following local treatment with curative intent, the PSA test is a good indicator of the presence of prostatic tissue, although it does not indicate its nature. In the large majority of cases of patients developing metastases after local treatment, a rise in serum PSA precedes evidence of secondaries with imaging by months to years.

Once the laboratory and imaging data (if indicated) together with other relevant information including comorbidities and social responsibilities have been evaluated, a considered decision has to be made about what to do—if anything at this stage. Importantly, *it is very seldom that a decision needs to be made immediately or in the next few weeks or even a month or so, but unproductive prevarication should be avoided* so patients and families can get on with their lives. As at the time of initial diagnosis, a second opinion is always an option as is taking an accompanying friend or relative to consultations to minimise the likelihood of missing or misinterpreting proffered information.

Considerations with further treatments: If further therapeutic intervention is required at this juncture, it is important to appreciate that all treatments have potentially significant unwanted effects (outlined below) and it is important to tailor their likelihood as closely as possible with each individual patient's priorities. There are increasing numbers of options becoming available based on the results of clinical trials so there is a suite from which to choose. These are discussed below.

Clinical trials: Because such studies are often very expensive and take time to provide clinically meaningful findings, of necessity most will have been commenced years before based on cutting-edge findings at that time so, if a patient wishes to have access to the very latest promising (though not yet proven or approved) treatment, he should consider participating in a trial.

Randomised Controlled Trials (**RCTs**) involve a trial preparation as one option with state-of-the-art treatment (which, in some instances, may be a placebo) as the other arm. Patients cannot "pick and choose" the arm to which they are allocated (usually computer-generated) and they need to commit to staying with the trial for its duration—unless something dramatic occurs to warrant withdrawing. In blinded studies, the treating staff and patients do not know the form of treatment allocated—although this certainly does not stop speculation.

Other benefits of trial participation include closer interactions with treating staff, particularly trial coordinators and study nurse(s), as well as progressive third-party scrutiny of patient welfare along with reviewing justification of continuing or terminating trials, by institutional ethics and other committees: these assessments are conducted with individual patients remaining anonymous, as they are in any publication(s) resulting from trials.

Chapter 6

First-line androgen deprivation therapy (ADT)

It is sobering to appreciate that, unlike men with clinically localised prostate cancer, most diagnosed with metastatic tumours die from their disease, with the large majority doing so within of 5 years of detection. Of those who die of noncancer causes, cardiovascular and cerebrovascular causes constitute the large majority [85,259,260].

Androgen deprivation therapy (ADT) continues to be the cornerstone treatment in the management of nonlocalised and metastatic prostate cancer, as well as being adjunctive to radiation treatment for localised disease, as mentioned previously (**Chapter 5** ADT with radiotherapy). Currently, all treatments for metastatic disease are palliative and are associated with side effects, so the quality of life aspects should be considered carefully, especially for men with significant other comorbidities and/or limited life expectancies. Increasingly, however, longer survivals are being reported by combining newer generation antiandrogens/androgen receptor (AR) signalling inhibitors and docetaxel chemotherapy (**Chapter 7** Taxanes) with first-line ADT at the time of metastatic disease detection [261,262].

Basal and luminal subtypes of prostate cancer are known to be present in metastatic tumours as well as primary cancers, Luminal metastatic tumours have overexpression of genes in the AR pathway compared with basal metastases in which involved genes predominantly affect cell cycle and cell division pathways. This has prompted the suggestion that those with the latter cell type would benefit more from chemotherapy approaches directed to small cell/neuroendocrine tumours rather than drugs specifically targeting the AR [263].

The promise of transcriptomic subtyping to guide prognostication in metastatic hormone-sensitive prostate cancer and selection of chemo-hormonal treatment was given a further boost by findings from an analysis of RNA signatures of prostate cancer material from the CHAARTED study. Patients with the luminal subtype had a poorer prognosis if they received ADT alone but benefitted significantly from a combination of ADT and docetaxel. Participants with the basal subtype with lower AR activity (and a higher risk with the Decipher genomic classifier) had a worse overall survival in this study, these findings

An Insider's Perspective of Prostate Cancer: Understanding Effects, Management Options and Consequences.
DOI: https://doi.org/10.1016/B978-0-443-18709-4.00013-4

indicating the differentiating role of biomarker classifications to potentially help in directing therapeutic approaches [264].

Androgen Receptor: The prostate is a male hormone (**androgen**, from the Greek "andr" meaning man) dependent gland during both normal development and in prostate cancer. In addition to the classical action of the AR which is to regulate gene transcriptional processes, it has a nongenomic role in signal transduction that involves modulation of cytoplasmic or cell membrane regulatory proteins [265].

The AR is indiscriminate in that it has the capacity to bind to all androgens but does so, particularly with the most common androgens testosterone (**T**) and its metabolite dihydrotestosterone (**DHT**), formed through the enzymic action of 5 alpha-reductase. T is produced predominantly by Leydig cells in the testicles—although some is also formed via dehydroepiandrosterone (**DHEA**) made by the zona reticularis, the innermost part of the adrenal cortex. (Both adrenal glands are situated above and overlap the uppermost part of their respective kidneys. Each has an outer adrenal cortex with three zones, glomerulosa, fasciculata, and reticularis, and an inner adrenal medulla).

The AR is complexed with chaperone proteins and located in the cytoplasm of androgen recipient cells. After ligand binding, mostly with DHT which has a higher binding affinity for the AR than T, the complex translocates to the cell nucleus to recognise DNA response elements in regulatory regions of androgen target genes. Interaction with epigenetic factors and other proteins follows resulting in transcriptionally active complexes to stimulate gene expression [266]. (The epigenome activates or inactivates [switches on and off] target genes). During prostatic development, T and DHT induce and facilitate growth via ARs in both epithelial and stromal cells [267]. However, this situation is changed in prostate cancer development and especially aggressive cancers with loss of ARs in stromal (connective tissue) cells and overexpression of ARs in malignant epithelial cells [268,269] which become androgen avid.

6.1 Hypogonadism (testosterone deficiency) with ageing

Late-onset hypogonadism is common as men age, the prevalence depending on the definition used. In the Baltimore Longitudinal Study of Ageing, hypogonadal testosterone levels were reported in approximately 20% of men over 60 years, 30% over 70 and 50% over 80 years of age [270]. Relating hypogonadism clinically based on symptoms that included poor morning erections, low sexual desire, erectile dysfunction, inability to perform vigorous activity, depression, fatigue, and limited physical vigour, a large European study found that late onset hypogonadism was present in between 3.1% and 7% in men younger than 70 years and 18.4% in those older than 70 years [271].

As circulating testosterone falls, oestrogen/estrogen (feminising/female hormone) levels in the blood rise by up to 40%. As well as a decline in testicular function, this change is due to a conversion to oestrogens (through a chemical process called aromatisation). Aromatisation affects fat deposits, common in older men, and male hormones including those produced in the adrenal glands (situated at the top of the kidneys). As a result, there are decreased DHT and increased oestrogen levels present in both the circulation and in prostatic epithelial cells with ageing [272]. Correspondingly, the relative relationship of oestrogens to androgens varies amongst racial groups paralleling differences in prostate cancer development.

6.2 Forms of androgen deprivation therapy

ADT was popularised following the 1941 publication by Huggins and Hodges who reported that prostate cancer regressed following **bilateral orchidectomy**/orchiectomy (surgical removal of both testicles) [273,274]. Subsequently, a variety of other treatments emerged to achieve castrate testosterone levels. However, bilateral orchidectomy remains the reference treatment despite being very largely replaced therapeutically by Luteinising Hormone Releasing Hormone (**LHRH**) or gonadotrophin hormone releasing hormone (**GnRH**) agonists and, to a lesser extent, LHRH or GnRH antagonists and antiandrogens. Previously both LHRH agonists (such as goserelin, Zoladex and leuprorelin acetate, Lucrin) and antagonists (such as abarelix, Plenaxis, and degarelix, Firmagon) were available only as injectables administered intermittently but recently, development of effective oral LHRH antagonist drugs (such as relugolix, Relumina) may increase acceptance compared with LHRH agonist and antagonist therapy administered by injection [275].

Following commencement of ADT, there is a prompt and significant response in terms of reduction of tumour volume in the large majority of patients; however, this effect is not permanent with tumour recurrence evident in 80% of cases within 12–36 months [276–280]. The clinical effect of ADT is considered to be primarily due to apoptosis (programmed cell death) of androgen sensitive luminal epithelial and prostate cancer epithelial cells [281,282] with changes also to stromal (connective tissue), smooth muscle and vascular endothelial cells occurring [283].

Because ADT is so important in the management of men with nonlocalised prostate cancer, base testosterone levels are relevant in patients considering ADT as those starting from a lower base could be expected to have a less pronounced benefit from this treatment than their eugonadal (normal testosterone-level) peers [284].

6.3 Bilateral orchidectomy (bilateral orchiectomy)

With bilateral orchidectomy (surgical castration), removal of testosterone-producing Leydig cells results in inactivation of a feedback loop to the hypothalamus (at the base of the brain immediately above the pituitary gland) (Fig. 6.1). As a result, the hypothalmus loses its stimulus to produce LHRH to activate the

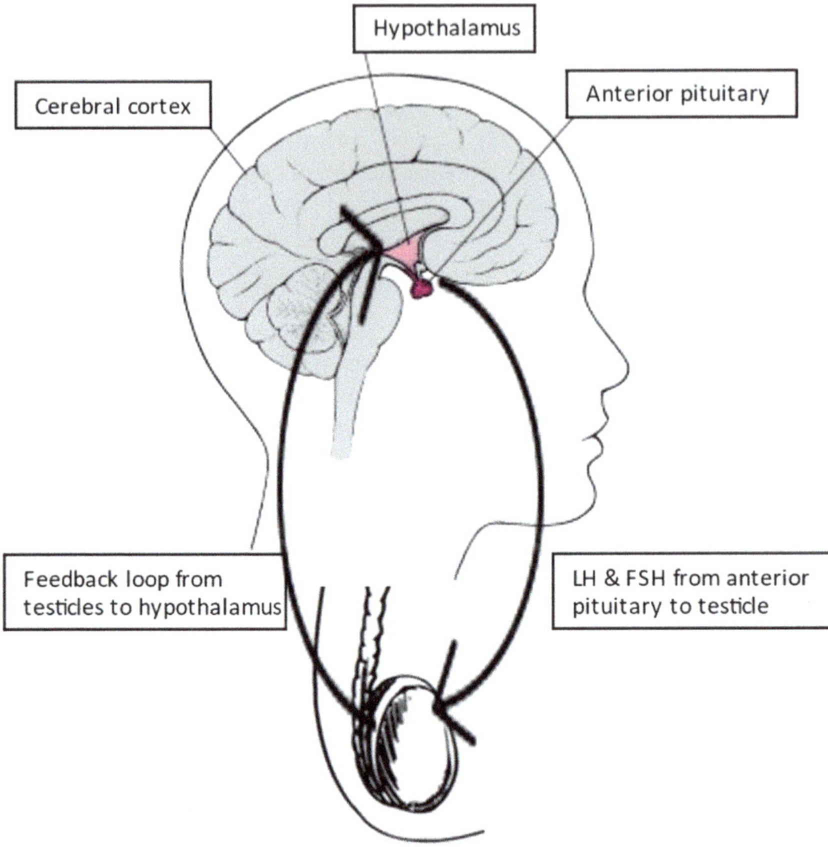

The pituitary gland or hypophysis is referred to as the master gland of the endocrine system. Situated at the base of the brain immediately below the hypothalamus with which it is intimately related anatomically and functionally, the pituitary has anterior and posterior parts with distinctly different functions.

The anterior pituitary produces hormones that travel to the thyroid, adrenals and testicles (as well as other structures) which in turn produce thyroxine, cortisol and testosterone, respectively, amongst other hormones. Regulation of these hormones is via feedback loops to the hypothalamus and anterior pituitary.

A hormone produced by the hypothalamus and released by the posterior pituitary, antidiuretic hormone or ADH, is vital in water resorption by the collecting ducts in the kidneys.

FIGURE 6.1 Normal male feedback loop to and from testicles.

anterior pituitary (sited immediately below) to produce Luteinising Hormone (**LH**)–destined via the bloodstream for Leydig cells to induce them to produce testosterone.

6.4 Oestrogens/estrogens

Understandably the irreversible prospect of surgical castration is not usually welcomed so, for many years, **oestrogens**/estrogens, mostly in the form of the synthetic preparation stilboestrol, were prescribed to suppress testosterone production. Oestrogens served to inhibit hypothalamic production of LHRH, as a result of this drug's induced positive feedback to the hypothalamus, and thus suppression of LH production by the pituitary gland at the base of the brain (Fig. 6.1). Stilboestrol also acted directly on prostate cancer cells via cell cycle arrest and the induction of apoptosis (programmed cell death) [285,286].

Oestrogen therapy, the commonest form of ADT for many years, is now rarely used as first-line hormonal treatment because of associated cardiovascular complications with oral administration, even in men without overt cardiovascular disease [287]. However, a dose–response relationship has been reported with one mg daily of diethylstilboestrol (with or without aspirin) stated to be comparable with bilateral orchidectomy in the treatment of advanced disease, but without increased cardiovascular complications [288]. In addition to oral oestrogens, intravenous oestrogens were given to men with bone metastases periodically in the past. Many patients indicated they could feel a painful effect in their bones for a short time after injection, attributed to targeted activity, with the sites of temporary discomfort corresponding with radiologically demonstrable bone secondaries.

An increased synthesis of coagulation factors, in particular Factor VII, was considered responsible for the increased rate of cardiovascular problems with oral oestrogens [289,290]. An appreciation of this information prompted a reconsideration of oestrogens, especially in the form of parenteral or transdermal [291] modes of delivery to avoid the "first past the liver" effect inevitable with oral administration [292]. (Molecules ingested and absorbed through the intestine go to the liver for processing before proceeding into the general circulation. Injections into muscle and topical delivery of medications by-pass the liver so the general circulation is accessed first). A further benefit advanced for the renewed interest in oestrogens is that, unlike bilateral orchidectomy and LHRH drugs, oestrogens do not induce osteoporosis [293] and possibly may even increase bone mineral density [294,295].

Oestrogen effects are commonly mediated by two receptors, oestrogen receptor α (**ERα**) and oestrogen receptor β (**ERβ**) which are members of a family of nuclear receptors that are often altered in prostate cancer. In the nonmalignant prostate, ERα is expressed predominantly in prostatic stromal cells with ERβ found mostly in basal-epithelial cells [296]. In prostate cancer, expression of ERα is associated with a high Gleason Score and a poor prognosis [297,298] with

ERβ reduced or absent [297,298]. ERα has been reported to trigger the tumour-promoting function of TMPRSS2-ERG (from fusion of the androgen-sensitive TMPRSS2 gene and the oncogenic transcription factor ERG), a common change in prostate cancer [299].

These findings spurred an interest in developing ERα antagonists for treating advanced prostate cancer along with ERβ agonists, called selective estrogen receptor modulators (**SERMs**) with experimental work indicating that ERβ agonists may have an anti-oncogenic role as well as being bone protective [300–304] but not affecting the development of gynaecomastia ("man-boobs") [305], common with ADT approach.

6.5 LHRH agonists and antagonists

LHRH agonists, or gonadotrophic releasing hormone (GnRH) agonists, target LHRH receptors in the pituitary, initially stimulating then blocking LH release and, as a result, Leydig cells in the testes are unable to produce testosterone. The initial testosterone production surge from initial LHRH receptor stimulation is called a "testosterone flare". LHRH agonist drugs, by blocking further LHRH release, then result in castrate circulating testosterone levels. By contrast, LHRH antagonists block LH production immediately and so there is not the "flare" problem of LHRH agonists so patients receiving LHRH antagonists do not require temporary additional "covering" testosterone suppressive therapy as do men administered LHRH agonist injections. LHRH receptors are not just limited to the hypothalamus, also being present on prostate tumour cells so LHRH agonists and antagonists have a direct effect on these cells [306–308] too.

6.6 Unwanted effects of LHRH agonists and antagonists

Cardiovascular disease is the commonest cause of death for men diagnosed and treated for prostate cancer with a considerable body of recent evidence indicating that LHRH antagonists are associated with a lower cardiac risk than LHRH agonists [309,310]. In addition to a low circulating testosterone being associated with an increased risk of cardiovascular morbidity and mortality [311,312], many patients considering the commencement of ADT have other risk factors for cardiovascular disease or have existing cardiovascular conditions. Consequently, not only should the relative risks of LHRH antagonists versus LHRH agonists be taken into account for these patients, but the timing of commencement for the palliative strategy that is ADT should be carefully considered in terms of overall likely survival duration and quality of life.

A further benefit of LHRH antagonists is that they prevent follicle-stimulating hormone (**FSH**) production via the anterior pituitary. This is relevant as, in addition to LHRH receptors, FSH receptors are also present on prostate cells with their expression increased in prostate cancer [313].

In addition to the unwanted effects of bilateral orchidectomy or surgical castration, there are additional side effects associated with LHRH treatments (Table 6.1).

TABLE 6.1 **Unwanted effects with bilateral orchidectomy and LHRH targeted medications.**

Unwanted or side effects	Bilateral orchidectomy	LHRH treatments
Hot flushes/flashes	+	+
Osteoporosis/osteopaenia (Loss of bone substance)	+	+
Loss of muscle mass	+	+
Metabolic syndrome	+	+
Cognitive changes	+	+
Reduced libido & impotence	+	+
Lowered haemoglobin (lowered red blood cell count)	+	+
Gynaecomastia ("man-boobs")	+	+
Flare reaction		+ (agonists only)
Arthralgia & myalgia (painful joints & muscles)		+
Dyspnoea (shortness of breath)		+
Peripheral oedema (Fluid accumulation in dependent parts)		+
Dizziness/lightheadedness/headaches		+
Nausea		+
Skin reactions		+

6.6.1 Hot flushes/hot flashes

Similar to women who experience hot flushes at menopause due to a reduction in oestrogen, men are prone to hot flushes with ADT due to testosterone suppression. Some patients become distressed and embarrassed with this not uncommon symptom which can be associated with profuse sweating. These episodic vasomotor changes can occur at any time without warning and are difficult to control therapeutically: selective serotonin reuptake inhibitors (**SSRIs**), such as venlaflaxine (Effexor), the anticonvulsant gabapentin, the progestogen medroxyprogesterone acetate, and oestrogen patches have been used with success for some patients [314,315].

6.6.2 Osteoporosis

Osteoporosis and osteopaenia/osteopenia (a less severe form of bone substance loss) are underdiagnosed in men with multiple contributing factors having been identified: these include physical inactivity, hypogonadism and ageing, smoking, excessive alcohol intake, and corticosteroid drugs. Men with osteoporosis are particularly prone to fractures of the neck of femur and vertebral bodies of the lumbar spine with the risk in prostate cancer patients [316,317] reported to be related to the duration of ADT with increased porosity of cortical bone and loss of trabeculae, evident by 12 months [318].

The incidence of hip fractures increases markedly after the age of 65 years [319], very often following seemingly low-impact trauma. A significant number never walk again with the associated mortality higher for men compared with women. In addition to shortening their height from vertebral body collapse, fractures of the lumbar vertebrae can cause nerve root compression and pain and, rarely, can involve the spinal cord tragically resulting in paraplegia [320–323]. However, prostate cancer secondary tumour expansion into the spinal canal has to be excluded as the cause of the paraplegia.

Vitamin D: Calcium and vitamin D supplementation (as indicated through monitoring levels of the vitamin D metabolite, calcifediol, also known as calcidiol, (25(OH)D)) are generally recommended, although the benefit in preventing fractures for men receiving ADT is uncertain. Dietary sources such as egg yolk, oily free-range fish, and fortified foods are beneficial with moderate exposure to ultraviolet light particularly very effective. Skin synthesis of vitamin D is self-regulated so that overdosing from sun exposure is prevented by the body's homeostatic regulatory system [324].

In addition to its important role with calcium in maintaining bone health, a recent study has indicated that vitamin D may have a beneficial role in enhancing survival for men with aggressive prostate cancer [325]. Vitamin D deficiency is surprisingly common in the general population in many parts of the world, particularly during the winter months with elderly and dark-skinned people living in colder climes especially susceptible.

Exercise: The value of exercise cannot be overemphasised in bone loss prevention. Combined impact loading and resistance exercise reduces bone loss in ADT patients and enhances musculoskeletal function. It should be commenced at the initiation of ADT to maximise benefit and, ideally, conducted under supervision [326–328].

Bisphosphonates & Denosumab: In addition, both bisphosphonates and the human monoclonal antibody denosumab (marketed as Prolia and Xgeva) are effective in stopping loss and increasing BMD in men receiving ADT. Bisphosphonates are analogues of pyrophosphate, a normal by-product of metabolism. They are absorbed effectively to hydroxyapatite, the crystalline form of calcium and phosphate in bone [329].

At sites of higher bone turnover, **osteoclasts** (cells that resorb bone) absorb bisphosphonates resulting in slowing of their bone-resorbing activity. Bisphosphonates are also reported to interfere with osteoclast formation, as well as resulting in premature death of these cells and consequently, favourably change the balance between **osteoblasts** (bone substance producing cells) and osteoclasts.

Osteoclast precursors (called pre-osteoclasts) express receptor activator of nuclear factor-kappa B (**RANK**) on their cell surfaces. RANK is activated by RANK ligand (**RANKL**) to enable preosteoclasts to become osteoclasts. The monoclonal antibody denosumab binds to RANKL to inhibit maturation of precursor pre-osteoclasts [330]. Thus, the mechanistic outcome from both bisphosphonates and denosumab is to slow bone resorption by osteoclasts.

A Cochrane Review published at the end of 2020 [331] concluded that the bisphosphonate zoledronic acid (marketed as Aclasta, Reclast, and Zometa) and denosumab, both given by injection, appeared to be effective in preventing skeletal-related events (complications related to bone secondaries and/or cancer treatment) in men with prostate cancer and bone metastases but these agents can also cause serious adverse events. These results align with those of current national and international guidelines for men with bone metastases from castration-resistant prostate cancer.

The bisphosphonate zoledronate/zoledronic acid is usually given intravenously as Zometa to prostate cancer patients, with its efficacy comparable with denosumab administered by subcutaneous injection. Both medications are well-tolerated with most patients not experiencing any side-effects; however, the most common that do occur with intravenous zoledronate (Zometa) are a temporary rise in temperature and mild flu-like symptoms, nausea, loss of appetite, and skin irritation/rash for a day or two. Similarly, most men receiving denosumab do not experience any side effects but the most common reported are also a rise in temperature and possibly chills, and skin irritations and rashes [332]. Hypocalcaemia can occur with both medications and may warrant calcium and vitamin D supplementation.

An extensive list of other possible side-effects is beyond the scope of this book but two warrant special mention. Worsening kidney function can occur with zoledronate so, as a precaution, patients with severe kidney disease should not receive intravenous bisphosphonates. A rare but devastating condition that is associated with both intravenous zoledronate and denosumab is osteonecrosis of the jaw (**ONJ**). In this condition, jaw pain is associated with an offensive breath, and exposed dead bone is visible in the mouth. ONJ usually follows dental work or an oral wound that has not healed: patients with bad dental hygiene are considered more at risk of developing this condition. As a precaution, it is advised that any outstanding dental work is undertaken and completed with the mouth healed before either of these treatments is commenced.

6.6.3 Loss of muscle mass

Loss of muscle mass, also called **sarcopaenia**/sarcopenia, is closely correlated with ageing. Understanding of the pathophysiology remains incomplete although exercise is one of the few strategies that improves muscle strength and functioning, minimising muscle loss and even reversing this process, including for men with advanced disease.

Exercise: The benefit of exercise in maintaining and reversing muscle wasting is often overlooked in prostate cancer treatment. Exercise is considered to have beneficial effects in terms of many important bodily processes which include, epigenetic effects on gene expression and DNA repair, oxidative stress and antioxidant pathways, immunity, chronic inflammation and prostaglandins, energy metabolism, and insulin resistance [333].

Both resistance training and aerobic (endurance) training provide benefits for patients with prostate cancer. Aerobic training predominantly improves cardiovascular and metabolic health which will be discussed shortly. Resistance exercise is the most effective therapy to increase muscle anabolism (metabolic "building-up") and decrease catabolism (metabolic breakdown) via the pathways mentioned above with mitochondria contributing to energy provision [334,335]. This ability to reverse muscle loss with resistance exercising regimens has been demonstrated conclusively in men receiving ADT [336,337] who also benefit in terms of associated reductions in anxiety and depression together with enhanced social connectedness [338].

Molecular basis: Muscle growth and size are stimulated via insulin and insulin growth factor-1 (**IGF-1**) via the PI3K/Akt/mTOR pathway. By contrast, FoxO1, FoxO3, and FoxO4 are members of a family of mostly nuclear transcription factors which, when activated, lead to muscle atrophy. However, when these FOXO members are phosphorylated by Akt, their transcription ability is thwarted enabling anabolic (building) activities in muscle cells to proceed via mTOR protein. Myostatin, produced by skeletal myocytes, along with other proteins (in particular, SMAD 2 and 3) interfere with IGF1-Akt signalling. mTOR signalling in turn inhibits SMAD 2/3.

Thus, at various molecular levels, there is normally a homeostatic balance exemplified by the anabolic (metabolic building) effects of the insulin/IGF—Akt/mTOR pathway contributing to muscle growth and the catabolic (breaking down) effects via Myostatin and SMADs opposing this action. This, however, is far from the complete story: a recent report indicates that activation of the JAK/STAT1 signalling pathway contributes to muscle weakness which can be partially prevented by JAK/STAT1 inhibitors [335]. Actions by nuclear factor κB (**NF-κB**), involved in immune modulation as well as cell survival and proliferation, and mitogen-activated protein kinase (**MAPK**) pathways influencing gene expression, cell division, metabolism, motility, survival, apoptosis, and differentiation amongst others are also recognised to play a role in the balance between muscle growth and development and muscle wasting [334].

Multisystem interactions: Muscle tissue is now recognised as an important endocrine organ influencing all other systems, organs, and tissues in the body including cancerous tumours [339]. When the muscle is activated (i.e., exercise) cytokines (molecules that affect other cells termed *myokines* when originating from muscle) are released into the systemic circulation. In addition, exercise increases blood pressure and body temperature as well as driving surges in immune cell numbers and activation. Exercise also increases blood flow through prostate tumours delivering these myokines and immune cells to the cancer cells with activated NKs and other T cells able to employ aerobic glycolysis to rapidly fulfill their important functions [52].

Alteration in levels of myokines IL-6, IL-15, IL-10, irisin, secreted protein acidic risk in cysteine (**SPARC**), myostatin, oncostatin M and decorin have variously been demonstrated to exert a direct inhibitory effect on cancer growth. Mechanisms include inhibiting proliferation, promoting apoptosis, inducing cell-cycle arrest, and inhibiting the epithermal transition to mesenchymal cells [340]. Thus, it is highly desirable for men with prostate cancer to maintain or increase their muscle mass and to activate it regularly to create and maintain a tumour-suppressive systemic environment.

Any exercise is better than none although supervised exercise with aerobic, resistance, and impact loading components is best to address adverse aspects of ADT physically and psychologically. *Resistance training is more effective in increasing muscle mass and strength: by comparison, maximum aerobic capacity is improved more by endurance training.* Furthermore, an added benefit of exercise and supervised exercise, in particular, is improved mental health especially those with the highest levels of psychological distress who are reported to improve most. Consequently, it is recommended that supervised exercise should be strongly recommended as a matter of routine for all men receiving ADT [93,341].

6.6.4 Gynaecomastia/gynecomastia

Gynaecomastia, also known as "man boobs," is a benign or noncancerous enlargement of one or both breasts: breast pain can occur as a further annoying accompaniment. Not all men who are castrated develop gynaecomastia. In their dissertation on history and the long-term consequences of castration in men, Wilson and Roehborn reported that some, but not all eunuchs, developed gynaecomastia but, in those who did in the Ottoman Court, the breasts became large and pendulous [342].

Again, there is background noise that needs to be considered with the presence of gynaecomastia, reported to be present in up to 60% of men aged 50–70 years [343,344]. In addition, a number of conditions and medications can cause gynaecomastia and these also need to be considered in the management of men with this condition [343,345,346]. Gynaecomastia has been reported to be associated with major psychological consequences that include depression, anxiety, and reduced self-esteem as well as adverse effects on quality of life [343,345–347].

The selective estrogen/oestrogen modulator (**SERM**) tamoxifen, which blocks the oestrogenic effect on breast tissue, is effective in management of gynaecomastia resulting from ADT for many men, with relatively low dose radiation to the breast area also an option [343]. Radiation can be given before ADT to reduce the number of ducts in the breast as well as in established cases. However, if gynaecomastia has been present for longer than a year, regression with these treatments is unlikely and subcutaneous mastectomy or liposuction may be required [343,348].

In those cases of gynaecomastia attributed to ADT, breast enlargement is considered to be due to oestrogen stimulation via ERβ receptors [349]. The hormonal imbalance responsible for gynaecomastia is thought to also adversely affect metabolic health with increased circulating oestrogen and reduced testosterone levels associated with obesity and type II diabetes [350].

6.6.5 Metabolic syndrome

Testosterone deficiency is associated with an increased risk of developing the metabolic syndrome [351–354]. The metabolic syndrome is defined as a cluster of three or more of the following and includes being treated for these conditions: excess abdominal weight, blood lipid (fat) abnormalities, high blood pressure, and an increased circulating blood glucose (hyperglycaemia) [351–354]. Central obesity (also called "belly fat") is common in older males in western countries with insulin resistance a frequent accompaniment.

Men with the metabolic syndrome are at greater risk of cardiovascular disease and of dying prematurely compared with those without it [356,357] together with a resulting higher risk of cardiovascular events, hyperglycaemia, and high circulating levels of insulin (via the PI3K/Akt/mTOR pathway). Hyperinsulinaemia also favours the growth of prostate cancer cells (see **Chapter 4**).

In addition to causing chronic hyperinsulinaemia, adipose (fat) tissue produces leptin, another potent driver of cancer progression. Through highly expressed receptors on cancer cells, leptin binding increases tumour cell proliferation, inhibits apoptosis (programmed cell death), promotes angiogenesis (vessel formation), and induces anticancer drug resistance.

Diet has an important part to play in the management of metabolic syndrome together with pharmacological management of triglyceride abnormalities, high blood pressure, and inducing patients to undertake exercise. In terms of diet to reduce and maintain weight, foods that have a low glycaemic index (**GI**), such as whole grains, dairy foods, legumes, and fruit, are preferable to those with a high GI such as white bread, biscuits, and cake. Alcohol, especially in the form of beer, is calorific so should be limited to maximise efficacy of any dietary strategy and, of course, for cardiovascular health, smoking should be avoided completely [314].

To achieve meaningful fat loss and ameliorate metabolic syndrome and cardiovascular disease risk, restriction of dietary energy intake is the most effective therapy. However, physical exercise is also helpful, in particular aerobic exercise involving large muscle groups activated over extended time periods such as with walking, jogging, cycling, swimming, etc. For fat loss, the amount of exercise needs to be 300 min per week or more. It is also essential to perform resistance training for at least two sessions per week to avoid the muscle and bone loss that accompany any caloric restriction diet.

As highlighted above, myokines are protein cytokines synthesised and secreted by skeletal myocytes (muscle cells) in response to muscle contraction [357]: adipokines are cytokines and proteins synthesised and secreted by adipose tissue. Both are recognised to be beneficial modulators of obesity, metabolic syndrome, and type II diabetes [358]. In addition to an autocrine function (affecting itself) in regulating muscle metabolism, myokines have a regulatory function on distant tissues such as bone and adipose tissue. Release of both myokines and adipokines (**Chapter 4**) with exercise is also beneficial for patients with a number of conditions other than, the metabolic syndrome with the rationale for its benefit in the management of metabolic syndrome continuing to mount, already being overwhelmingly persuasive.

6.6.6 Cognition

According to the Oxford Dictionary, cognition is the mental action or process of acquiring knowledge and understanding through thought, experience, and the senses. Cognitive decline is a common and insidious feature of ageing with many (often unappreciated) drugs contributing. Amongst these are medications with anticholinergic properties, which include those used for the treatment of depression, Parkinsons Disease, and overactive bladder. In addition, there are others with weaker but definite anticholinergic properties [359–361]. The list of medications that can adversely affect cognition is extensive and includes commonly prescribed drugs such as digoxin and frusemide, which have weak anticholinergic properties [359].

Polypharmacy is common. In Australia in 2017–2018, there were 1012 prescriptions for every 100 men aged 45–54, progressively increasing through the decades to 5918 for every 100 for those 85 and over [362]. Together with nonprescription medications, the cumulative effect of drugs on cognition is potentially highly significant. Consequently, a considerable number of men commencing ADT have less than a "clean sheet" in terms of drug intake so indications for any further treatments that may compound an already suboptimal cognitive status deserve careful consideration.

Cancer-related cognitive impairment is increasingly being recognised as a serious and common problem during and after treatment. In the medical scientific literature, a number of studies have reported that up to 50% of men receiving ADT with LHRH agonists have objective evidence of adverse cognitive changes in tests of verbal memory, visual information processing, and executive functioning [363–368]. In the only randomised controlled trial, none of the men in the 'no treatment' or in a community control reference group had significant cognitive deterioration at 6 or12 months, in particular in relation to complex information processing. The cognitive defects documented in the 50% of men affected adversely were not insignificant as they were of a magnitude comparable with sleep-deprivation or mild inebriation [364–368]. These changes, often not recognised by the patients themselves, have been attributed to a specific genetic polymorphism [367,368].

Functional LHRH/GnRH receptors in the brain are not only present in the human pituitary and hypothalamus but also in the cortex and hippocampus [369,370], one of the first parts of the brain to suffer damage in Alzheimer's Disease [369–371]. Early symptoms of Alzheimer's Disease include changes reported in association with ADT which, together with epidemiological data, have prompted concern that ADT predisposes to subsequent development of Alzheimer's Disease in a proportion of men [372,373] with similar concerns also expressed from findings in relation to anticholinergic drugs [374].

Thus, in terms of cognitive decline, avoiding the "death by a thousand cuts" effect that drugs that adversely affect brain function can induce should be a desirable strategy—whenever possible. *Consequently, together noncognitive adverse effects, the decision to commence ADT with LHRH analogues should be a carefully considered decision and not taken lightly.*

Chapter 7

Castrate resistant prostate cancer (CRPC)

Bilateral orchidectomy/castration, LHRH (GnRH) agonists/antagonists and oestrogen/estrogen, directly and indirectly, starve androgen receptors (ARs) of testosterone and dihydrotestosterone. Also used as first-line androgen deprivation therapies (ADTs) in the past, alone or in combination, are nonsteroidal antiandrogens flutamide, nilutamide, bicalutamide, and the steroidal antiandrogen cyproterone acetate. However, these are not strong enough by themselves and/or have unacceptable toxicity profiles for inducing castrate circulating testosterone levels to be recommended as first-line ADT medications, although bicalutamide is used not infrequently short-term with LHRH agonists to offset the initial flare reaction.

7.1 Emergence of castrate-resistant prostate cancer

ADT with the above treatments is effective in causing shrinkage of prostate cancer secondaries in most patients with this effect sometimes being dramatic for those with widespread metastases. However, this is but a temporary respite with recurrence occurring in 80% of patients within 12–36 months of commencing androgen deprivation treatment [276–280,375,376].

Castrate resistant prostate cancer (CRPC) has been subclassified. Nonmetastatic castrate-resistant prostate cancer (**nmCRPC**)—means that the PSA is rising despite first-line ADT but secondary cancers are not evident with standard imaging—although PSMA PET imaging has "pushed the boundary" as a more sensitive indicator in that it permits identification of metastases that were otherwise previously "invisible" [103]. Metastatic castration resistant (**mCRPC**) disease is the same as nmCRPC—except that metastases *are* evident with imaging. The nmCRPC group of patients is more heterogeneous as *some* men with nmCRPC have elevated PSAs with (not yet visible) secondaries that only become apparent with imaging subsequently, suggesting that their metastases do not express the marker used for the imaging or that the metastases are progressing at a slower rate [377].

An Insider's Perspective of Prostate Cancer: Understanding Effects, Management Options and Consequences.
DOI: https://doi.org/10.1016/B978-0-443-18709-4.00006-7

7.2 Diagnosis of CRPC

In the presence of a castrate level of circulating testosterone, the first clinical manifestation of "escape" from what is regarded as first-line ADT is usually a rising PSA. As stated above, the definition of the PSA level at which this is said to occur varies, but the most common level is the same as following radical prostatectomy, namely, 0.2 ng/mL. This biochemical escape, previously referred to as "hormone escape" is the first evidence of the clinical state of castrate resistance and may precede imaging evidence by a considerable period of time, varying with the aggressiveness of the "escaped" cancer.

The reason for changing the name of this phase in the so-called natural history of the disease from "hormone escape" to CRPC is that those cancer cells surviving first-line ADT suppression continue to progress via hormones and other molecules binding to ARs. There is a change in AR-coactivators in the cell cytoplasm leading to enhanced AR-mediated gene transcription, postulated to promote progression toward CRPC with the above process serving as a molecular switch [378].

In an environment of a circulating castrate testosterone level, ARs on cancer cells increase in number and configuration so that they are not only available to bind to any androgen molecule in the immediate vicinity but, because of post-translational alterations to the AR, they are also able to bind to molecules other than androgens which include other hormones and some drugs. These modified receptors are called promiscuous ARs [379–383]. Added to these changes in AR receptivity is the capacity of some cancer cells to produce molecules, including their own androgens, that bind to their ARs as a form of intracrine or self-stimulation.

Molecular guidance: There is increasingly a role for molecular findings to indicate the most effective management approaches. In addition to being able to predict significant cancers at diagnosis and time to biochemical escape more accurately, molecular profiling enables targets for both approved drugs and others, in the process of development and preclinical assessment, to be identifiable. Men with mCRPC often are debilitated from their prostate cancer, various comorbidities, and treatments and are approaching the end of their prostate cancer journey, so a "try it and see approach" with further options has limited appeal and warrants careful consideration, especially as further treatments almost always bring with them significant additional side effects.

Splice variants, active mRNAs that result from cutting and resealing of RNA transcripts at exon (coding segment)—intron (noncoding segment) junctions, are associated with the emergence of CRPC. The splice variant protein AR-V7 is rarely present in primary prostate tumours but is detected in three quarters of cases of prostate cancer cells with tumour progression following first-line ADT, its prevalence increasing further after abiraterone or enzalutamide administration (see below). The presence of the AR-V7 splice variant protein is associated with a reduced response to AR targeting drugs with correspondingly worse outcomes in terms of PSA response and overall survival, with the converse also applying [384,385].

Other genomic alterations in circulating tumour cells from "AR-V7 negative mCRPC men" and associated with poor outcomes from enzalutamide/abiraterone, include copy number alterations in genes linked to lineage plasticity and epigenetic signalling, DNA repair, AR, TP53/RB1, PTEN, and WNT pathways [386]. A recent study reported a high concordance for DNA repair gene alterations in primary prostate cancer, circulating DNA and metastases [387].

Molecular information is increasingly becoming available to guide management strategies to maximise efficacy and minimise unproductive therapeutic forays. This knowledge for an individual patient may result in a different or additional strategy in management. (Further molecular changes relevant in therapy selection are indicated below).

7.3 Earlier strategies

For many years, bilateral adrenalectomy (surgical removal of both adrenal glands) was a treatment option, performed to remove the source of adrenal androgens, in men whose cancers were no longer controlled by their primary ADT which, in those days, was either bilateral orchidectomy or oral oestrogen therapy, usually stilboestrol. The two big drawbacks to bilateral adrenalectomy were that the procedure was a big (open) abdominal operation for advanced cancer patients with limited life expectancies and there was no way of predicting who would or would not benefit. For these reasons, bilateral adrenalectomy was not a universally popular option.

A less invasive alternative was to target the pituitary gland (or hypophysis) via a trans-sphenoidal approach (via the nose and the sphenoidal air sinus), by surgically dislocating and removing the pituitary gland from its stalk (or pedicle) and thus disconnecting it from the hypothalamus immediately above and with which its function was inextricably related (see Fig. 7.1). A variation on the theme was to destroy the pituitary by introducing radiotherapy as an yttrium implant to the pituitary fossa (or depression) in the bone at the base of the skull just below the pituitary gland (also called the sella turcica as it was considered by early anatomists to appear similar to a Turk's saddle), via the same trans-sphenoidal route (Fig. 7.1). Although a relatively minor intervention, there was always a risk of leakage of cerebrospinal fluid postoperatively following both these trans-sphenoidal procedures, which usually settled spontaneously, but an associated risk of meningitis was always present.

After both trans-sphenoidal destruction of the pituitary gland and bilateral adrenalectomy, hormone replacement therapy was required. Because the pituitary is the site from which many hormones are released to target a number of vital organs, this required a much more extensive prescription list

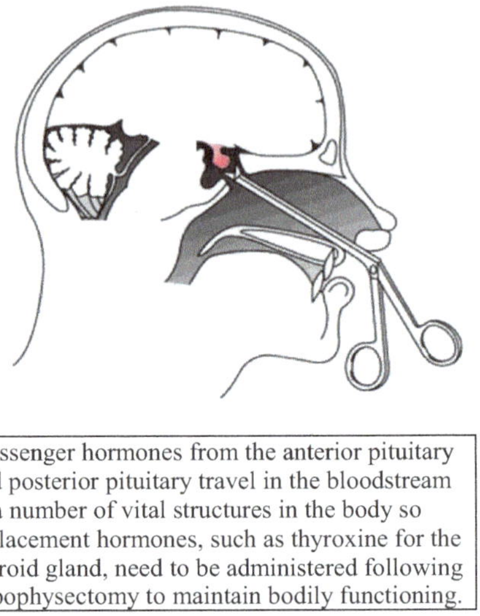

FIGURE 7.1 Trans-sphenoidal hypophysectomy (pituitary gland removal via sphenoidal air sinus).

posthypophysectomy, which included cortisol, thyroxine, and an antidiuretic hormone analogue, than following bilateral adrenalectomy with cortisol alone usually needed.

7.4 Newer generation antiandrogens: androgen receptor signalling inhibitors

More recently, a new generation of drugs was introduced, called AR signalling Inhibitors, that are much more effective than first-line LHRH and earlier generation nonsteroidal antiandrogen drugs mentioned above, but approximately between 20% and 40% of CRPC patients who receive enzalutamide or abiraterone acetate do not have a PSA response [385].

7.4.1 Enzalutamide (Xtandi)

Enzalutamide is a nonsteroidal newer-generation antiandrogen that has a much greater affinity for binding to the AR than any of its first-generation antiandrogen predecessors. By binding to AR, enzalutamide competitively inhibits testosterone and dihydrotestosterone from binding to it, preventing changes normally induced by testosterone-AR binding that, following nuclear translocation of the bound complex from the cytoplasm, normally leads to DNA transcription (transcribing DNA to RNA) by the cell. Thus, enzalutamide is effective at more than one level in preventing proliferation of those cells with ARs to which it binds.

However, as with all chemotherapeutic drugs in prostate cancer, duration of the beneficial effects of enzalutamide is limited, being related to the timing of commencement of treatment (below) and the ability of cancer cells to adapt. Following AR binding by enzalutamide, some prostate cancer cells side-step this therapy to obtain their energy through cortisol binding with glucocorticoid receptors: in these cells 11β-HSD2, a protein that inactivates cortisol and is typically expressed by tumours, becomes inhibited so that cortisol-glucocorticoid receptor binding occurs thus providing an alternative source of fuel [388] so those benefitting "escaping" cancer cells can continue on their rampaging way.

In a randomised, double-blind, placebo-controlled, phase 3 trial of enzalutamide or placebo undertaken at 254 study sites world-wide (PROSPER Trial), men with nonmetastatic CRPC in the enzalutamide arm experienced longer metastasis-free survival than did those who received placebo, the primary end-point, with lower pain levels and high health-related quality of life, secondary end-points [389]. Participants recruited to both arms of the trial had asymptomatic CRPC and a PSA doubling time of 10 months or shorter, despite first-line ADT which was continued throughout the study [389].

In a phase 3 trial, patients *previously treated by the chemotherapeutic drug docetaxel* were randomised to enzalutamide or placebo: participants in both groups continued first-generation ADT medications. Enzalutamide was continued until progression of disease became evident at a median duration of 8.3 months, compared with 3 months for the placebo patients. The median overall survival was significantly longer for those in the enzalutamide treatment group at 18.4 months compared with 13.6 months for the placebo group participants [390]. Unwanted effects were common with asthenia/fatigue experienced followed by back pain in 50.6%–26.4%, respectively [390].

Two further large randomised multinational trials were published recently *but* in patients with metastatic *hormone-sensitive* prostate cancer. In the first, patients were assigned to receive testosterone suppression plus either open-label enzalutamide or a standard nonsteroidal antiandrogen therapy [391,392]. In the second published, later in the same year, participants were assigned to enzalutamide or placebo, plus ADT, stratified by disease volume and prior docetaxel chemotherapy [392]. The primary endpoint for the first trial was overall survival and, for the second study, radiographic progression-free survival. There was a statistically significant benefit for men randomised to enzalutamide plus first-line ADT in terms of overall and progression-free survival in both trials.

Consequently, the addition of enzalutamide is now approved and recommended to be given with first-line ADT to men with *metastatic hormone sensitive* prostate cancer following diagnosis. As with first-line ADT, prostate cancer suppression with enzalutamide is but temporary due to the development of drug resistance. Enzalutamide interacts with a number of commonly prescribed medications [393] with very common side effects including fatigue, nausea, hot flushes, diarrhoea, hypertension (high blood pressure), gynaecomastia (man boobs), dry skin, pruritis (itchiness), restless legs syndrome, and falls. In addition, enzalutamide has been reported to upregulate the long noncoding RNA p21 leading to neuroendocrine tumour dedifferentiation, a highly aggressive form of prostate cancer [394,395].

Enzalutamide crosses the blood–brain barrier with accompanying central nervous system side effects considered to be linked to enzalutamide-AR complex binding in brain cells with headaches and cognitive disorders also reported with this medication. To address this issue objectively in terms of cognitive changes, the results of a trial currently underway, as flagged by Lange et al., are eagerly awaited [396].

7.4.2 Apalutamide (Erlyand)

Another androgen signalling inhibitor, apalutamide, became available recently for clinical use [397]. Also, unlike normal androgen-AR binding in which the complex translocates from the cytoplasm to the nucleus where it activates AR-responsive genes, the apalutamide-AR complex remains in the cytoplasm (in a manner similar to that which occurs with enzalutamide).

A randomised study of patients with *metastatic, castration-sensitive prostate cancer* reported that those who received apalutamide in addition to their first-line ADT experienced a significantly longer radiographic progression-free survival and overall survival compared with participants who received ADT and placebo. At a median follow-up of 22.7 months, the radiographic free survivals were 68.2% and 47.5%, respectively, and for overall survival at 24 months were 82.4% and 73.5%, respectively. Grades 3 or 4 adverse events were 42.2% in the apalutamide group and 40.8% in the placebo group [398].

In a randomised trial of men with *nonmetastatic CRPC at high risk for metastases developing by virtue of a PSA doubling time of 10 months or shorter*, participants in both arms continued their standard first-generation ADT. A planned analysis after 378 events found that the median metastases-free survival, the primary endpoint, was 40.5 months for men in the apalutamide group compared with 16.2 months for those in the placebo group [399].

Fatigue and low platelet, white and red cell counts are common with apalutamide and may be problematic. Similarly, apalutamide combined with ADT has been shown to be superior to ADT plus placebo in delaying the onset of CRPC and slowing decline in health-related quality of life in men with *metastatic castration sensitive* prostate cancer [400]. Other common side effects experienced by patients taking apalutamide include hot flushes, nausea, diarrhoea, increased blood pressure, dizziness/falling, and peripheral oedema [401]. Emergency medical help should be sought if signs of an allergic reaction develop, such as difficulty breathing, swelling of the face, lips, or tongue.

7.4.3 Darolutamide (Nubeqa)

Another family member darolutamide is structurally different to other AR signalling inhibitors. This drug is reported to target both wild-type and mutated ARs in the cytoplasm inhibiting AR nuclear translocation [402] but, unlike other

AR signalling inhibitors, darolutamide appears to result in a considerably lower level of blood–brain penetration [403,404]. As a result, it is speculated that this medication is likely to have less of an adverse effect on cognition compared with other second-generation antiandrogens.

A higher percentage of men with *nonmetastatic castration-resistant prostate cancer* randomised to receive darolutamide rather than placebo remained alive at 3 years in a phase 3 trial. First-line ADT was continued by participants in both groups. A significant benefit for all secondary end points, including time to first symptomatic skeletal event and time to first cytotoxic chemotherapy, were also reported for men administered darolutamide. Those randomised to receive daralutamide also had a 31% lower risk of death compared with those men who received placebo in addition to their first-line ADT [405].

Many of the unwanted effects acknowledged with enzalutamide and apalutamide are reported with daralutamide treatment. Fatigue, rashes, and pain in extremities are also noted by some patients [406]. Because direct comparative data to guide treatment decisions was not available, a meta-analysis was undertaken to compare efficacy and side effects for enzalutamide, apalutamide, and darolutamide. The authors concluded that the first two drugs appeared to be more efficacious than darolutamide in the treatment of nonmetastatic prostate cancer but that, of the three, darolutamide appeared to have the most favourable tolerability profile [407].

7.4.4 Abiraterone acetate (Zytiga)

By contrast Abiraterone acetate, another AR signalling inhibitor with a different mode of action, became available at about the same time as enzalutamide. Normally, adrenocortical trophic hormone (**ACTH**) produced by the anterior pituitary following directions from the hypothalamus, produces pregnenolone in the adrenal cortex which, in turn leads to the production of the salt retaining hormone aldosterone by one pathway and dehydroepiandrosterone (**DHEA**) by another which then leads via androstenedione ultimately to testosterone and oestradiol (Fig. 7.2).

By blocking the pathway from pregnenolone to 17-hydroxypregnenolone, there is a shunt in pregnenolone metabolism by abiraterone resulting in increased aldosterone production along with decreased production of cortisol or hydrocortisone.

The mineralocorticoid hormone aldosterone is produced by the zona glomerulosa (outermost part) of the adrenal cortex, sited immediately above, and overlapping the uppermost parts of the kidneys. Aldosterone acts primarily on the distal tubules and collecting ducts of the kidneys to result in sodium retention at the expense of potassium loss. The resulting imbalance between sodium and potassium can lead to water retention and contribute to blood pressure elevation with low potassium predisposing to heart dysrhythmias so, a mineralocorticoid antagonist may be required to be added to the drug regimen of men receiving abiraterone acetate in some cases.

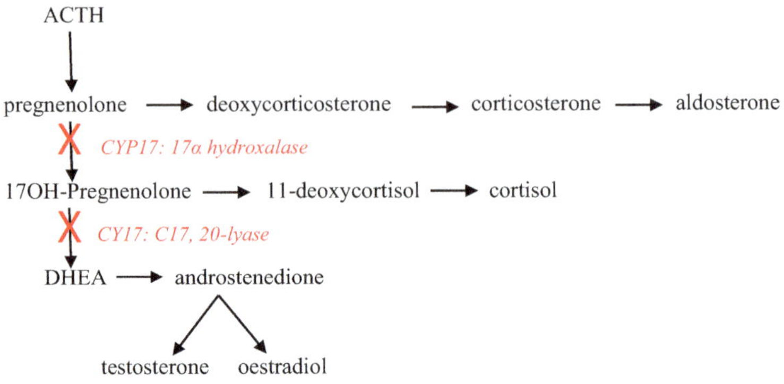

FIGURE 7.2 Abiraterone mechanism of action.

On the other hand, because production of cortisol (the body's major glucocorticoid) is suppressed by abiraterone, synthetic derivatives such as prednisone, methylprednisolone, or dexamethasone are essential to compensate. Cortisol has many functions which, amongst others, include the breakdown of fat and protein that are then converted by the liver to glucose in a process called gluconeogenesis.

Clinical studies: Two large, randomised trials of *metastatic castration-resistant* prostate cancer patients who had not had and who had previously received docetaxel chemotherapy, in which participants were randomly assigned prednisone with either abiraterone acetate or placebo, found a clear benefit for men in the abiraterone group. For participants who had not had previous chemotherapy, the time to commencing cytotoxic chemotherapy, opiate use for cancer-related pain, PSA progression, and decline in performance status favoured those in the abiraterone group [408]. Postchemotherapy, the median overall survival for those men who received abiraterone was longer than for participants who were allocated placebo with similar beneficial differences for time to PSA progression and for radiologic progression-free survival [409].

A meta-analysis of the addition of abiraterone acetate plus prednisone or prednisolone to standard first-line ADT in men with *metastatic hormone-sensitive disease* who had not received docetaxel found a highly significant 38% reduction in the risk of death and a highly significant 55% reduction of clinical/radiological progression-free survival for the first and second-line ADT combination. They concluded that abiraterone was an alternative to docetaxel for men with metastatic hormone-sensitive prostate cancer prescribed first-line ADT for the first time [261].

7.4.5 Comparisons and combinations

As mentioned in the introduction, detailed lists of side (or unwanted) effects of medications are beyond the scope of this book and for these the reader is referred elsewhere. However, in comparing enzalutamide and abiraterone, a susceptibility to potential cardiovascular side effects with both drugs has received particular attention. In their review of cardiovascular toxicity from prospective studies, Iacovelli et al. (2017) indicated that abiraterone significantly increased the risk of both cardiac toxicity and hypertension (high blood pressure) compared with enzalutamide which only increased the risk of hypertension [410]. More recently, Lu-Yao et al. reported that in men with pre-existing cardiovascular comorbidities, there was an increased mortality and hospitalisation risk following enzalutamide *and* abiraterone acetate [411,412].

The oncological activity of enzalutamide is reduced after abiraterone, docetaxel, and even more after both. This lack of sequential benefit appears to prevail for all members of the new generation antiandrogens with the greatest oncological advantage expected if these new-generation agents are administered earlier in the natural history of prostate cancer than later [413].

A recent analysis of results from a study of men with *chemotherapy-naïve metastatic* CRPC treated with apalutamide and abiraterone acetate plus prednisone together compared with abiraterone acetate plus prednisone, demonstrated a benefit for the combination for radiographic progression-free survival but did not show an improvement in overall survival [414]. Similarly, there was no significant difference in time to initiation of cytotoxic chemotherapy, time to chronic opioid use or time to pain progression between the groups: serious adverse effects were recorded for 40% and 37% of participants, in the apalutamide–abiraterone combination and abiraterone groups, respectively [414,415].

7.4.6 Androgen receptor signalling inhibitors and cognition

The introduction of newer generation ADT to men already chemically castrated also raises concerns with respect to cognition, with no objective assessments using validated and internationally accepted instruments undertaken and published to the author's knowledge. Although these drugs pass through the blood–brain barrier, it would appear that there is a difference between them in ability to do so, particularly with daralutamide [416]. Whether ARs in the central nervous system change in response to castration effects from first-line ADTs is also an unresolved question. However, what is certain is that rigorous cognitive studies are required especially since there is a trend to introduce these medications increasingly early.

7.4.7 Androgen receptor signalling inhibitors in hormone-sensitive prostate cancer

In a recent monograph on enzalutamide in men with *metastatic hormone-sensitive* prostate cancer, the invited author's concluding comment was that clinical factors and patient preference will determine whether an individual with *metastatic hormone-sensitive* prostate cancer should receive ADT combined with docetaxel, or enzalutamide, or abiraterone, or apalutamide, or whether he will have ADT alone [417]. This comment is in keeping with recent recommendations in American and European guidelines [418–420]. However, for an individual patient, these guidelines may need to be tempered by emerging evidence from genomic studies indicating that men with predominantly luminal tumours respond better to AR signalling inhibitor therapies than those whose cancers are predominantly basal in cell type who are more likely to respond to chemotherapy targeting DNA repair pathways [263].

7.5 Systemic radiation therapy—Theranostics

Although prostate cancers spread to both lymph nodes and bone, the latter usually dominates. In advanced disease, bone secondaries can be debilitating both in terms of pain and disability from fractures, hence the attractiveness of a systemic form of radiation that targets bone and particularly bone secondaries. The use of systemic radiotherapy (also referred to as "liquid radiation therapy") in medicine has a long history, being best known in the context of thyroid disorders and radioactive iodine (I^{131}). The term "**Theranostics**" is commonly used to describe demonstration of tumour by radio-isotope imaging followed by targeted radioisotope therapy to treat tumour deposits expressing the common marker.

7.5.1 Strontium and samarium radioisotopes

An isolated bone secondary may be treated with external beam radiotherapy to relieve pain, which it can do quite effectively, but widespread skeletal metastases are much less suitable for this so-called "spot-welding" approach. Bone-seeking radioisotopic compounds, in particular the beta-emitting strontium^{89} and the gamma and beta-emitting samarium^{153}-ethylenediaminetetramethylene (**EDTMP**), were used in the past but enjoyed limited success due to their associated bone marrow suppressive effects. Another problem with strontium^{89} was its long half-life of approximately 50 days compared with 46 h for samarium^{153}. Both these agents concentrate in bone metastases with little dose effect in soft tissues. Following intravenous administration, maximal effect on pain was usually experienced between 2 and 4 weeks [421–425].

7.5.2 Radium-223 chloride (Alpharadin)

Alpharadin is an alpha-emitting radionuclide of radium ($^{223}RaCl_2$). It is preferentially absorbed by osteoblasts into bone as a calcium analogue and has proven clinically safe and well-tolerated for the palliation of bone metastases in patients with prostate and other cancers [426]. Alpharadin administration has been reported to improve survival in patients with castration-resistant prostate cancer with symptomatic bone metastases for up to 3 years following first injection [427]. Its therapeutic limitations are that it does not target soft tissue metastases (although there may be a small bystander effect as alpharadin is an alpha emitter) and there is a higher risk of fracture following this treatment. Consequently, it is strongly advised that a bone health-protecting agent (see Osteoporosis section in **Chapter 4**) is taken by patients for whom alpharadin is prescribed [428].

7.5.3 PSMA radionuclide treatment

Prostate-specific membrane antigen (PSMA) expression increases progressively in higher-grade cancers, metastatic, and castration-resistant disease [429–432]. Although first thought to be prostate-specific [432,433] PSMA expression has been identified in cells of the small intestine, kidneys, and salivary glands, albeit at levels 100–1000-fold less than in prostate cancer [100]. PSMA is expressed by the majority of prostate cancer secondaries and their associated new blood vessels, but not on normal vasculature [99].

^{68}Ga PSMA PET CT scanning is used to detect prostate cancer metastases (see section staging prostate cancer in Chapter 4) with, subsequently, PSMA tagged via another isotope as radionuclide therapy. This is usually in the form of lutetium-177 [^{177}Lu[Lu-PSMA-617 (Novartis) [434,435], a small molecule inhibitor that binds with high affinity to PSMA and is a beta emitter, with other isotopes such as alpha emitters able to be used as well. Lutetium-177 PSMA radionuclide therapy has been demonstrated to result in disappearance of metastases, assessed by subsequent ^{68}Ga PSMA PET CT imaging, increase survival, and to relieve unrelenting bone pain in men with CRPC [436].

Other targets for theranostics are currently being pursued. Amongst a number of promising candidates are Gastrin-releasing peptide receptors which are expressed on the surface of a prostate and other cancers. These receptors are able to be targeted with synthetic ligands in a similar manner with different isotopes attached for imaging and therapy [437], respectively.

A recently published phase 2 trial involving men with metastatic CRPC, for whom carbazitaxel chemotherapy (see below) was considered the next appropriate form of treatment, were randomised to carbazitaxel or Lu-PSMA-617. With a primary end point of a PSA reduction of at least 50% from baseline, there was a highly significant difference between the two arms in favour of men in the Lu-PSMA-617 group with grades 3–4 adverse events in 33% of men in the Lu-PSMA-617 arm compared with 53% in the carbazitaxel arm [438]. However, others have reported a lower incidence of adverse events for chemotherapy-naïve patients with metastatic CRPC [439]. The published benefits of Lu-PSMA-617 in patients with advanced CRPC continue to accrue with yet another large multicentre study also very recently reporting prolonged image-based progression-free survival and overall survival [440].

Although the utility of lutetium-177 [^{177}Lu[Lu-PSMA-617 therapeutically is attractive with advocates suggesting it should be used earlier in disease progression before ADT, not all cancer cells express PSMA so a negative imaging ^{68}Ga PSMA PET CT scan following lutetium-177 [^{177}Lu[Lu-PSMA-617 therapy may be misleading in that secondaries *not* expressing PSMA may not be appreciated with diagnostic or post-treatment monitoring ^{68}Ga PSMA PET CT scans. As a result, although lutetium-177 [^{177}Lu[Lu-PSMA-617 [434,435] is a valuable new therapy in metastatic disease, it is not considered curative, at least as monotherapy. As inferred above, a proportion of metastases do not respond to PSMA labelled lutetium-177 with suboptimal PSMA target expression, inadequate radiation dose, and biological tumour resistant mechanisms proffered as explanations for failure.

7.6 Chemotherapy

The introduction of new chemotherapy drug treatments almost invariably follows clinical trials of men with advanced disease. If the drug being trialled does show a difference in delaying disease progression and/or in delaying death, the time intervals for determining efficacy are faster in CRPC patients so the trials can be considerably shorter to show a change than if the drugs in question were introduced earlier in the natural history of prostate cancer. In almost all of these trials, first-line ADT is continued with psychological and quality of life factors now established as essential inclusions for analysis in these studies along with, most recently, addition of distress evaluation.

Unfortunately, however, when chemotherapy is effective for patients with advanced disease, the respite is only short. To the casual observer, it may seem that the cancer is always a step or two ahead in that when it does return due to persisting and replicating surviving cells, it does so in an even more aggressive form requiring stronger and usually more toxic treatments that are added to continuing ADT. Consequently, it is pertinent to keep asking with each step "are further treatments justified"—is the addition of the "new" treatment making the overall therapy worse than the disease?

Until relatively recently, there was little that could be offered for patients with castrate-resistant disease (or hormone-escape as it was then called) in terms of treatments to reduce tumour burden: as a result, life expectancy for these patients was limited. However, with the many advances now based on an ever-increasing understanding of the molecular biology of prostate cancer, therapeutic efficacy continues to advance with survival periods becoming longer and longer so that, even for men with aggressive tumours, a diagnosis of prostate cancer becomes more of a chronic condition rather than an imminent death sentence.

With these advances, the real challenges are maintaining or improving the quality of life of all those undertaking the prostate cancer journey and at a reasonable financial cost. Consequently, the vital importance of clinical trials in which not just tumour effect from the treatment is scrutinised but also detailed assessment of the effects experienced by participants in terms of psychological changes and other side-effects are evaluated as all trial medications, including placebos (intentionally formulated inactive medications), have side effects above and beyond those from the drugs already being taken.

7.6.1 Estramustine phosphate (Emcyt/Estracyt) and ketoconazole (Nizoral)

Prostate cancer has generally been regarded as a chemo-resistant cancer with rates of clinical responses to various cytotoxic drugs, alone and in combination, disappointing historically. A combination of oestrogen to target metastases combined with a cytotoxic nitrogen mustard, known as estramustine phosphate, was introduced in the 1970s but was limited in its effect because of its side-effect profile [441].

The fungicide ketoconazole is a *nonselective* inhibitor of 17α-hydroxylase/17,20 lyase (CYP17A1) and continues to be used, albeit to a limited degree, as a second-line therapy for CRPC. Ketoconazole has clinical efficacy although a survival benefit has not been shown. It has a big advantage of being cheap, but liver toxicity has limited its use in the past: however, some consider this side-effect to have been overstated and that it can be avoided if not minimised with careful monitoring so that ketaconazole may still have a place for patients who do not respond to or are unable to tolerate other treatment options for medical or financial reasons [442]. However, its successor abiraterone acetate, which is a *selective* 17α-hydroxylase/17,20 lyase (CYP17A1) inhibitor (see newer generation antiandrogens: AR Signalling Inhibitors in this chapter above), has virtually relegated ketoconazole to the past.

7.6.2 Mitoxantrone (Novantrone)

Following a randomised study of men with symptomatic castrate resistant disease, those who received mitoxantrone and the glucocorticoid prednisone had a significant reduction in their need for analgesia and for a longer time compared with patients in the control group who received prednisone alone. A subsequent trial of mitoxantrone with hydrocortisone found that, although there was no survival benefit, there was an overall improvement in quality of life and a duration of analgesic response of 8 months on average for participants in the treatment group compared with men in the control arm [276,443]. As a result, the combination of mitoxantrone plus glucocorticoid became the accepted standard treatment for men with symptomatic castrate-resistant disease and remains a consideration for palliation today when further options are no longer available.

7.6.3 Taxanes

The taxane docetaxel (Taxotere) was the first cytotoxic drug to demonstrate a survival benefit in CRPC patients and, along with its second-line alternative, carbazitaxel (Jevtana), remains a commonly prescribed drug. Taxanes work by interfering with the ability of cells to depolymerize the microtubule component of their cytoskeletons during normal mitosis, thus inducing a lethal cell cycle arrest at the G2/M phase. (The cytoskeleton is a microscopic network of protein filaments and tubules, of which microtubules are but one component in the cytoplasm of living cells, that gives them shape).

Although docetaxel and carbazetaxol result in only a modest survival benefit of several months for patients with advanced disease [444,445], two large trials (CHAARTED and STAMPEDE) reported that combining docetaxel therapy with first-line ADT early for patients with recently diagnosed *metastatic hormone sensitive* tumours, the median overall survival time was increased significantly [446,447]. Consequently, for men with demonstrable metastases at diagnosis commencement of docetaxel chemotherapy with first-line ADT has been advocated at the time of starting long-term maintenance first-line ADT [446,447].

However, the addition of docetaxel does come at a cost with up to 50% of men receiving ADT and docetaxel reported to experience grade 3 (of 4) toxicity so predicting who will and will not respond would be very helpful. A recent report indicates that the long noncoding RNA grow arrest-specific 5 (GAS5) has been proposed to serve as a predictive marker with lower levels of expression associated with resistance to docetaxel in men with CRPC [448].

Docetaxel is known to cause severe nausea and vomiting with thromboembolic problems (blockage of a blood vessel(s) by a clot dislodged from one site and moving to another) also problematic for a number of patients with all aspects of quality of life having been reported to deteriorate during its use [449,450]. This is of particular concern in the elderly with prostate cancer, together with the potential for adverse cognitive effects from this drug, based on reports of its use with other cancers [451–453]. As the combination of ADT together with docetaxel has been stated to be a standard of care for de novo (new) high-volume metastatic prostate cancer, studies are underway to offset docetaxel's toxicity, yet maintain or even enhance its efficacy, by combining it sequentially with other treatments such as 177Lu-PSMA [454].

7.6.4 A recent change in treatment selection

As with most cancers, until very recently, prostate cancer treatment was based on the results of trials and experience, using a "rule of thumb" contemporarily agreed approach. However recently, with a greater understanding of the molecular processes involved with aggressive cancer replication and progression as well as an ability to detect such changes relatively inexpensively, strategies to target identifiable molecular vulnerabilities in patients' cancer cells have begun to be integrated into guidelines and recommendations as well as practice [455,456].

7.6.5 Olaparib (Lynparza) and talazoparib (Talzenna)

Aggressive cancers cycle more rapidly and, in so doing, the fidelity of their progeny suffers due to aberrations in DNA transcription, cell cycle regulation, and DNA repair, aided by induced impaired immune detection and cytotoxicity. As a result, there is a higher rate of cell death and turn-over, than for normal or less aggressive cancer cells. Drugs directed to impairing those repair processes that remain operating effectively in cancer cells have been developed. Two such drugs are Olaparib and Talazoparib which inhibit the enzyme poly adenosine polyphosphate ribose polymerase (**PARP**) to kill cancer cells: patients with BRCA 1 and 2 mutations in their cancers have an increased reliance on PARP to repair their DNA to permit them to continue dividing [457].

A randomised, phase 3 study evaluated Olaparib in patients with *metastatic CRPC* who had disease progression while taking enzalutamide or abiraterone. The cohort with at least one alteration of BRCA1, BRCA2 or ATM genes identified in their tumours, had a significantly longer progression-free survival of 7.4 months and a median overall survival of 18.5 months compared with 3.6 months and 15.1 months in the control group (with mutations in other genes) [458]. Multiple guidelines and consensus panels now recommend germline testing for men with metastatic prostate cancer [459], particularly those with metastatic CRPC (see below).

Talazoparib also inhibits PARP enzymes impairing the ability of cancer cells with BRCA1 and BRCA2 mutations to repair DNA damage, thus interfering with cancer cell replication. This treatment is reported to target prostate cancer cells, but spare normal cells in metastatic castration-resistant disease patients [460]. Common side effects reported are fatigue, anaemia, nausea, neutropaenia (reduced numbers of neutrophils), headache, thrombocytopaenia, anorexia vomiting, hair loss, and diarrhoea [461], amongst others.

7.6.6 Ipataserib (Kevzara) and sabizabulin

Ipataserib acts by blocking the PI3K/Akt signalling pathway, activated by loss of the tumour suppression gene PTEN, evident in up to 50% of metastatic, castration-resistant prostate cancers (mCRPC) [43]. Based on the premise that patients with dysregulation of both AR and P13K/Akt signalling pathways would benefit from a combination of the androgen signalling inhibitor abiraterone acetate and ipataserib, a multinational randomised-controlled trial was undertaken of untreated men with asymptomatic or mildly symptomatic metastatic CRPC.

At a median follow-up of 19 months, there was a significant benefit in terms of progression-free survival for 47% of participants with demonstrable PTEN loss who received ipataserib, abiraterone acetate and prednisone group at 18.5 months compared with 16.5 months for those administered placebo, abiraterone acetate, and prednisone. The overall results for progression free survival for men with *and* without PTEN loss detected were 19.2 months and 16.6 months, respectively. Grade 3 or higher adverse events were registered in 70% of participants in the ipataserib-abiraterone group and 39% in the placebo-abiraterone group leading to 21% and 5%, respectively, discontinuing their participation [462]. This trial is ongoing.

Another drug under investigation in men with CRPC is sabizabulin. Sabizabulin interferes with AR transport to inhibit microtubular polymerisation and thus disrupt formation of the cytoskeleton [463,464].

7.7 Immunotherapy

As indicated in the section on Immunological Competence (**Chapter 3**), the concept of the body healing itself is an attractive one for many reasons with initiation of modern immunotherapy attributed to Coley's published findings at the beginning of the twentieth century [465]. Following dramatic responses to some advanced cancers with immune checkpoint inhibitor treatments, in particular melanoma but also to a limited extent some other malignancies, interest in immunotherapies in general has been re-awakened with the host's overall immunological status rather than the tumour itself stated to be more important in predicting the potential efficacy of immunotherapy [466].

However, to effect cell killing, immune cells need to be robust and able to recognise and be in direct contact with cancer cells. As mentioned above aggressive tumours are able to prevent this interaction by a variety of mechanisms [467]. Unfortunately, to date, prostate cancer has not been demonstrated to be an immuno-responsive cancer, although some notable exceptions have been observed indicating that effective immunotherapy is a promise yet to be realised.

Many contrived immunotherapies are at-risk of targeting normal tissues as well as tumours in an autoimmune response because target antigens are often shared. Unlike theranostic radioligand treatments in which the duration of radiation-induced cell killing is more predictable, this is less so with immune-induced cell killing.

7.7.1 Immune checkpoint inhibitors (ICIs)

As indicated above, normally the immune system fulfills its important role of recognising and destroying abnormal cells without harming normal ones. In achieving this function, it is helped by immune checkpoints, molecules on relevant immune cells that need to be activated or inactivated for the immune response to proceed. An important checkpoint is between the programmed cell death ligand-1 (**PDL-1**) and its receptor on T lymphocytes. Another checkpoint inhibitor is the cytotoxic T lymphocyte-associated protein 4 (**CTLA-4**). Immune checkpoint inhibitor antibodies, such as pembrolizumab, nivolumab, and ipilimumab (below), have been developed and are used in selected patients with advanced prostate cancer. (The "mab" suffix indicates the monoclonal antibody status).

Significant side effects have been reported with immune checkpoint inhibitors (**ICIs**) with T cell activation leading beyond tumour-killing to involve normal tissues. These adverse effects range from rashes to, rarely, fatal conditions [467,468]: up to 30% of patients treated with the PD-1 and/or CTLA-4 inhibitors have been reported to experience grade 3 or 4 immune-related events [469].

7.7.2 Pembrolizumab (Keytruda) & nivolumab (Opdivo)

Programmed cell death ligands, or binding molecules, bind to PD-1 receptors on lymphocytes. PD-L1 protein expression can be induced on a cell as determined appropriate by that cell. When a PD-1 ligand binds to the PD-1 receptor on an immune cell, there is immune down-regulation with a reduction in T cell activity. A minority of prostate cancer cells upregulate the expression of PDL-1 and PDL-2 cell surface proteins that activate PD-1 receptors [470,471] on T-cells giving these cancer cells a survival advantage.

Pembrolizumab is a humanised murine (mouse) antibody of the immunoglobulin (IgG) family that inhibits normal interaction between PDL-1, and its receptor sited on the surface of activated T lymphocytes. By blocking binding of PDL-1 and its lymphocyte receptor, the interposing action of pembrolizumab is responsible for resulting in T cell-mediated cell death of those prostate cancer cells expressing PDL-1 by removing their "immune-privilege" status - but also runs the risk of causing unwanted autoimmune-induced inflammatory reactions in normal tissues. These potentially very serious pembrolizumab-induced side effects occur mostly in lung, a number of endocrine (hormone-producing) organs, kidney, liver, and colon.

Because patients with high levels of microsatellite instability (**MSI**) and deficient mismatch repair (**dMMR**) in their tumours are more likely to have a durable response after receiving checkpoint inhibitors [472,473], such as pembrolizumab, assays to establish MSI-H/dMMR are considered warranted before commencing this form of treatment. (Microsatellites are repeated sequences of DNA present in all people. MSI is the predisposition to mutation resulting from impaired DNA mismatch repair (**MMR**): MMR is the process that corrects spontaneously occurring errors during DNA replication. MMR proteins form a complex that bind to the mismatched segments of DNA, excise the error and insert the correct sequence).

Nivolumab is another checkpoint inhibitor. It is a human IgG monoclonal antibody that binds to the PD-1 receptor on T cells to prevent binding of PD-1 ligands on cancer cells to their receptor on T lymphocytes.

7.7.3 Ipilimumab (Yervoy)

Another approach is to target CTA-4. The cytotoxic T lymphocyte-associated protein 4 (CTLA-4) is an intracellular protein in resting T cells which, after suitable molecular inducement, translocates to the cell surface to lead to the arrest of T cell proliferation and activation. Ipilimumab and tremelimumab (an AstraZeneca trial drug) are human monoclonal antibodies that effectively block this action.

Studies with other cancers, notably with melanoma, showed promise albeit with not inconsiderable side effects in some patient such as enterocolitis and hepatitis amongst others which usually, but not always, could be controlled by corticosteroids without jeopardising the desired immunosuppressive effect [474].

In a multicentre, phase 3 trial, men with CRPC and at least one bone secondary who had progressed following docetaxel were randomised to bone-directed radiotherapy followed by ipilimumab or placebo. The primary end-point of overall survival for the two groups was 11.2 months and 10 months, respectively, the statistical difference being insignificant. There were grades 3–4 adverse events in 101 (26%) participants in the ipilimumab group compared with 11 (3%) of patients in the placebo group: 4 men died as a result of toxic effects from the study drug or placebo [475].

A trial in which combination of antibodies targeting both PDL-1 with Nivolumab and CTLA-4 with Ipilimumab recently found that, for mCRPC patients treated prechemotherapy and postchemotherapy (with 50% in each group), the objective response rates were 25% and 10% with overall survivals of 19 months and 15.2 months, respectively [476]. Interestingly, two participants in each group had complete responses.

7.7.4 Other types of immunotherapy trialled in prostate cancer

Chimeric antigen receptor (CAR) T cell therapy is another form of contrived treatment that, although showing great promise from studies with lymphoma and leukaemia, has yet to have this potential realised in prostate cancer with a significant stumbling block considered to be the heterogeneity (diversity) of aggressive prostate cancer cells and their antigens. CAR T cells are created by isolating T cells which are modified to have synthetic receptors (**CARs**) on their surfaces which enable the CAR T cells to recognise and attach to the target antigen. These CAR T cells are then expanded in the laboratory hundreds of million-fold before being ready to inject into patients. Antigen selection has been an issue in solid cancers with studies having been undertaken with prostate stem cell antigen (**PSCA**) [477] and PSMA [478].

Dendritic cell (DC) immunotherapy for prostate cancer has an established track-record with a number of trials having been undertaken. Interest was kindled by a publication using monocyte derived DCs (**MoDCs**) and PSMA peptides [479] in 1996. Although responses were obtained in this and subsequent MoDC-PSMA studies, as well as with the use of harvested secondary cancers as the source of antigen, responses were modest, with the exception of a very few participants [480–482]. Not all patients are eligible for a specific DC treatment because of tissue type limitations, this having been a factor limiting applicability. The best-known MoDC trial is the Sipuleucel-T (Provenge™) study in which prostatic acid phosphatase (**PAcP**) is the target antigen. Although initial reports with Sipuleucel-T were promising, experience with this expensive vaccine over many years has proved disappointing [483].

DC immunotherapy is attractive as it is almost always very well-tolerated resulting in only a minority of patients experiencing any side-effects at all. This strategy is designed to augment the immune repertoire complementing and enhancing, rather than detracting from, overall immune competence. In the past 20 years, much has been learned in optimising DC isolation and expansion [484] as well as T cell activation so a potential resurgence of this mode of treatment is expected in the next few years as important, recently discovered basic science principles are evaluated in clinical trials.

Chapter 8

Holistic medical management

8.1 Survivorship

An emerging area, particularly relevant for men with prostate cancer, is survivorship. Survivorship is much more than just existing longer. Domains include health promotion and advocacy, shared management, vigilance, personal agency, care co-ordination, and evidence-based survivorship interventions. In a recent systematic review of reviews, exercise and psychosocial interventions stood out as being effective in particularly improving survivorship outcomes [91]. These domains provide guidance for policymakers, clinicians, community, and consumers on what is essential for change in prostate cancer survivorship outcomes [91,92]. www.pcfa.org.au/awareness/survivorship-essentials-framework/.

However, it is one thing to recognise a need for patient support services but another to provide them. In a survey of 13 conventional cancer care supportive interventions between 2019 and 2020 in Australia, 265 cancer organisations were invited to participate. The criterion for selection was provision of at least one cancer-directed treatment or conventional cancer care interventions to cancer survivors. Of the 123 responses, psychological therapy (90%), self-care guidance (82%), exercise (77%), healthy eating (69%), and lymphoedema services (69%) were reported to be most common with fatigue management (51%) and pelvic health guidance [32] less often provided. Services offering massage, return-to-work, cognitive therapy, sleep hygiene, and leisure guidance were described as under-represented (<31%) [485].

Prostate cancer nurses fulfill a most important role in helping patients and their loved ones address their various concerns. Patients generally relate well to Prostate Cancer Nurses who, through mutual agreement and usually with more time than medical staff, are able to delve into issues of particular relevance for individual patients. Concerns at diagnosis and pretreatment include uncertainty of prognosis and fear of the unknown, worry, and anxiety about treatment options, appreciating how side-effects are likely to influence the quality of life, together with individually relevant social issues and responsibilities. By working closely and facilitating contact with medical specialist services and organisations such as the Cancer Council, Beyond Blue, PCFA (Australia), and other support groups internationally (see **Addendum**), Prostate Cancer Nurses are able to

An Insider's Perspective of Prostate Cancer: Understanding Effects, Management Options and Consequences.
DOI: https://doi.org/10.1016/B978-0-443-18709-4.00011-0

provide great solace and invaluable support for those in need throughout their duration of medical management.

Despite increasing efforts to improve the welfare of prostate cancer survivors, it is evident that much needs to be done. An emerging area in cancer treatment is using music as therapy. Benefits extolled include reducing anxiety, helping in the treatment of depression, easing acute and chronic pain, and alleviating fatigue [486]. There is also a need to address cardiovascular and cerebrovascular health in prostate cancer patients reflected by noncancer causes of death exceeding those from cancer with these diseases dominating [487–491].

Exacerbations of distress are inescapable for many men throughout the course of their disease. Chronic stress is reported to promote cancer development and progression through a number of mechanisms [492]. These include, for example, a stress-induced chronic inflammatory response leading to a decline in immune surveillance via the stress hormone noradrenaline (also called and norepinephrine). Binding to increased numbers of beta-adrenergic receptors expressed on myeloid-derived suppressor cells has been reported to lead to and promote an immunosuppressed tumour environment [493]. Consequently, strategies to reduce unnecessary stressful situations are important for patient welfare for much more than just psychological reasons.

In the development of prostate cancer, and especially in the emergence of its more aggressive forms, there is a commonality with a number of factors that increase the likelihood of men developing cardiovascular conditions [494]. In addition, a number of prostate cancer treatments contribute to cardiovascular pathologies [495,496] so there is considerable potential for improvement on many fronts with the first steps being recognition of the causative relationship together with appreciation of the magnitude of the issue and strategic options that may be employed to improve cardiovascular morbidity.

8.2 Palliative Care

Unfortunately, when chemotherapy is effective for patients with advanced disease, the respite is almost invariably short. To the casual observer, it may seem that the cancer is always a step or two ahead in that when it returns due to replication of surviving cells, it does so at an even faster rate and in a more aggressive form requiring stronger and usually more toxic treatments: these are almost invariably added to continuing androgen deprivation and other therapies. Consequently, it is pertinent to keep asking at each step "are further treatments justified—will this added treatment be worse than the disease"?

In the continuing evolution of medicine, the specialty of Palliative Care (**PC**) has developed from being one of caring for people at the very end of their lives to a specialty dedicated to helping people with a diagnosis expected to lead to their demise. Unfortunately, there are major differences in PC provision across the globe [497] but, even when PC services are available, many people consult a PC physician and PC team far too late, sometimes just before the patient finally succumbs! In addition, the recent preoccupation with voluntary assisted dying across much of the western world has been a distraction that has not helped the public's appreciation of what invaluable benefits PC services can offer patients.

The World Health Organisation defines PC as an approach that improves the quality of life of patients and their families facing the problems associated with life-threatening illnesses, through the prevention and relief of suffering by means of early identification, impeccable assessment, treatment of pain and other problems, physical, psychosocial, and spiritual [498].

Thus, the modus operandi of the PC team is to achieve and maintain the best quality of life for patients and their families, relieve pain, treat coexisting ailments and problems of a psychosocial and spiritual nature. Their scope of expertise is wide and interacts with many other specialties and services, with a focus on maximising for the prostate cancer patient what *he* considers to be important to him for the rest of his life. Indeed, some men regard a diagnosis of an aggressive prostate cancer as an opportunity to take stock and map out the future in a manner that otherwise never would have been undertaken. It is not so much the normal process of death itself that is so often feared but the dying due to its potential unpleasantness—which can be largely avoided. An excellent recently published book by Emeritus Professor Teik Oh AM dealing with this topic that can be recommended is "**An Extraordinary Journey: What Matters at the End of Life**."

Patients are advised to consider consulting a PC physician once a diagnosis of incurable prostate cancer is known. It is only by getting to know their patients and doing so early that PC team members can most effectively aid patients in achieving their individual goals. Emphases do vary but universal patient concerns include being able to pursue activities of interest with loved ones, remaining content and lucid and, of course, sleeping well and avoiding pain. Wherever possible, care is delivered in the environment of the patient's choice. PC physicians have a close working relationship with pain management specialists to ensure that any pain endured is managed optimally so that nobody lingers with unrelenting or distressing pain.

In their excellent document "Improving supportive and PC for adults with cancer," the National Institute for Health and Care Excellence (**NICE**) from the United Kingdom [499] state in their guidance foreword that it aims for people with cancer to be well informed, cared for and supported from ***before*** formal diagnosis onwards (author's emphasis). Their recommendations include the following:

- patient and carer involvement,
- psychological, social, and spiritual support services,
- general and specialist PC services,
- rehabilitation services,
- complementary therapy services,
- services for families and carers, and
- workforce development.

Chapter 9

Summary and conclusions—thoughts on the future

The future for patients with clinically significant prostate cancer is increasingly bright as, even for those diagnosed with advanced, aggressive disease, the condition continues to become more of a chronic one than an imminent death sentence. Continuing improvements in healthcare generally translate into more men living longer lives well, with clinically significant prostate cancer emerging as just another nuisance affliction bedevilling them in their later years. Because these patients are prone to harbouring a wide range of comorbidities, it is expected that less invasive and less disruptive forms of intervention such as focal therapies will increase in popularity and that prostate cancer treatments will be implemented with even a greater awareness of each individual patient's social and psychological circumstances than has been the case previously.

A greater emphasis on quality of life in management choices is inevitable as it is not living (or, in many cases merely existing) longer that is important. Prostate cancer is an affliction of older men who innately lack much of the energy of their youth and so many treatments add to this problem such that fatigue is a compounding and an under-recognised problem. Consequently, when choosing treatment strategies, maintenance therapies that increase fatigue and have side effects present day in day out should be distinguished from those with unwanted effects that are limited to only a short period of time. It is so easy for additional treatments prescribed with the well-intentioned objective of correcting aberrations identified with tests to result in unwittingly hastening these patients' inexorable physical and mental decline.

In addition to fatigue associated with ageing, loneliness, and boredom are blights adversely affecting the lives of many elderly men exacerbated, in this context, by a diagnosis and treatment of clinically significant prostate cancer. Participation in self-help groups (see **Addendum**) and exercise are recognised as invaluable, yet still overlooked and neglected, treatment options. As indicated earlier, any exercise is better than none, although supervised exercise with aerobic, resistance, and impact loading components is shown to be best for improving both physical and psychological patient status at all stages of prostate cancer and its treatment [500]. If there were a drug that was anywhere nearly as

An Insider's Perspective of Prostate Cancer: Understanding Effects, Management Options and Consequences.
DOI: https://doi.org/10.1016/B978-0-443-18709-4.00010-9

beneficial as exercise for patients with all stages of cancer, it would be lauded and prescribed widely.

The trend to fewer men being diagnosed with low-risk diseases is expected to continue due to better selection. As a result of promotion of emerging tests such as the **STHLM3** (Stockholm 3) and, more specifically, molecular profiling of liquid biopsies being integrated with PSA, mpMRI, and radioisotope imaging with PET CT, it is expected that the numbers with low-risk disease proceeding to biopsies and treatment will reduce even further in the near future.

Genetic profiling of biopsied prostatic tissue is advancing rapidly and will lead to better prediction of tumour behaviour [501] with the tailoring of treatment options appropriate to each individual's findings. The speed and costs of genetic profiling are continually reducing and are already starting to impact on tumour differentiation and treatment rationales. As a "follow on," it will be interesting to observe the comparably rapid advances in gene editing making inroads into already established prostate cancers and, ultimately, targeting at-risk men by preventive strategies. The challenge then, as always, will be to marry findings with available and relevant options for each individual's circumstances, priorities, and wishes.

The success being reported with lutetium-177 radioisotope treatment in men with advanced prostate cancers [434,435] is likely to lead to greater use of this and other isotopes in treating those with earlier disease using targets other than PSMA also being explored and evaluated. Together with radioisotope labelled imaging for more accurate disease staging to provide better localisation of primary and secondary tumours, this "self-fulfilling Theranostics prophecy" is expected to have an expanding role.

It will be interesting to see whether the renewed interest in immunotherapies and the promise they are bringing to patients with other cancers extends to men with prostate cancer. Certainly, researching the various forms of immunotherapy and how they might become harnessed and integrated with other prostate cancer therapies is proceeding apace, with those most promising likely to be fast-tracked to clinical trials quickly.

Health demands from ageing populations throughout the developed world exacerbated by unexpected scourges such as COVID 19 are impacting as never before on the escalating costs of delivering healthcare. Whether the main provider is the private sector or, as it appears to be happening increasingly, the public sector, the many and varied strategies emerging to improve the prognosis and welfare of patients all come at a cost with the money having to be found somewhere. Consequently, the importance of a strong health economics case is likely to become an essential requisite in any submission for a change in treatment or equipment so that cost analyses and projections will be a component of published studies in high-ranking journals increasingly in the future.

Glossary

ACE Inhibitors angiotensin-converting enzyme inhibitors.

Adenocarcinomas are cancers that arise from the epithelial (surface lining) cell component (as opposed to rarer sarcomas that arise from stromal or connective tissue components).

Adenosine triphosphate (ATP) consists of an adenosine molecule bonded to three phosphate groups. Breakage of one phosphate linkage (to form adenosine diphosphate, ADP) provides energy for a range of processes.

Adrenoreceptors (AdrRs)/adrenergic receptors membrane-bound proteins that mediate peripheral actions of noradrenaline/norepinephrine and adrenaline/epinephrine. There are two main types, α and β together with subtypes.

Adipokine Cytokine hormones made by adipose (fat) tissue (such as leptin and adiponectin).

Adrenal glands The two adrenal glands are situated above and overlap the uppermost part of their respective kidneys. Each has an outer adrenal cortex with three zones, glomerulosa, fasciculata, and reticularis, and an inner adrenal medulla.

ADT/ first-line androgen deprivation therapy involves the removal of testicular androgen production by either surgical or medical castration.

Anabolic means building and is used usually in reference to cells or tissues, for example, anabolic steroids and muscle. (The opposite is **catabolic** or breaking down).

Androgens male sex hormones, the best known of which is testosterone.

Androgen receptors in the cytoplasm bind with androgens such as testosterone, dihydrotestosterone, and dehydroepiandrosterone.

Angiogenesis is the process of blood vessel formation.

Antigen is a substance that induces an immune response.

Antigen Presenting Cells (APCs) are specialised cells that capture antigens, transport them to secondary lymphoid organs, and display them for recognition by lymphocytes.

Apoptosis is a naturally occurring process by which a cell is directed to programmed cell death.

Arthralgia painful joints.

Arthritis inflammation of joints.

AS (Active Surveillance) a management strategy for men diagnosed with low-risk prostate cancer considered unlikely to progress. It involves a prospective monitoring program that includes interval biopsies and an intention to change to a management with curative intent if parameters change to indicate a change in risk category.

ASTRO The American Society of Therapeutic Radiation and Oncology.

Atypical small acinar proliferation of uncertain significance foci of atypical glands that have features suspicious for but not diagnostic of cancer although they may and do accompany definite malignant changes.

Autocrine a secreted substance acting on itself.

Autophagy is a form of cell death in which cells consume adjacent cells of the same lineage, often as a stress response.

Benign tumours/tumors are abnormal masses of tissue that remain localised to the part of the body in which they arise.

B lymphocytes (also called B cells) are cells of the acquired immune system which, when activated by an antigen undergo clonal proliferation and maturation into an antibody producing **plasma cell**.

Benign tumours/tumours are abnormal masses of tissue that remain localised to the part of the body in which they arise. They can become very large as in the case of benign prostatic hyperplasia (BPH). By contrast, malignant or cancerous tumours have a propensity to invade surrounding tissues *and* to form secondary tumours in other sites in the body separated from the primary lesion by normal tissue.

Bilateral orchidectomy/orchiectomy is the surgical removal of both testicles.

Biochemical failure (BCF) refers to a rising PSA having reached a designated level with blood testing following treatment.

Bisphosphonates are a class of drugs that resembles pyrophosphate but are not readily hydrolysed. They interfere with the action of osteoclasts to prevent bone reabsorption.

BPH is the abbreviation for Benign Prostatic Hyperplasia which is an increase in size due to an increase in the numbers of constituent cells. BPH develops in the transition zone of the prostate.

CAR T Cell Therapy CAR T cells are created by isolating T cells which are modified to have synthetic receptors (chimeric antigen receptors) on their surfaces which enable the CAR T cells to recognise and attach to the target antigen.

Castrate Resistant Prostate Cancer (CRPC) Prostate cancers that are progressing clinically, no longer being constrained by first first-line androgen deprivation therapy (ADT).

CTCs circulating tumour cells.

Catabolic means breaking down.

CD is the abbreviation used universally to characterise cell surface molecules. CD refers to classification determinant or cluster designation or cluster of differentiation.

Chromatin is the DNA-protein complex in cell nuclei.

Cochrane Foundation, Cochrane Collaboration, and Cochrane Reviews The book "Effectiveness and Efficiency: Random Reflections on Health Services" by Archibald (Archie) Cochrane MBE, published in the early 1970s, laid the foundations for the Cochrane Collaboration. In this book, the lack of reliable evidence for many commonly accepted health care interventions was highlighted and criticised. Archie Cochrane's advocacy for the use of randomised trials (RCTs) and systematic reviews to make medicine more effective and efficient eventually led to the creation of the eponymously named Cochrane Foundation, founded in 1993 five years after Archie Cochrane's death.

Although Cochrane is a registered not-for-profit organisation with its headquarters in England, members and supporters come from more than 220 countries worldwide and consist of researchers, health professionals, patients, careers, and people passionate about improving health outcomes for everyone, everywhere. Cochrane's global, independent network gathers, and summarises the best evidence from research. Cochrane Reviews are highly regarded globally for their rigorous analyses of studies for guiding evidenced-based management.

Comorbidities are other medical conditions present.

chromothripsis the process in which chromosomes are fragmented and then reassembled in ways that promote growth.

Codon is a sequence of three nucleotides which that together forms a unit of genetic code.

CTLA-4 is the abbreviation for the cytotoxic T lymphocyte–associated protein 4.

CT scan (CAT scan) Computerised axial tomography are X-rays reconfigured by a computer to show sectional cuts of the body.

Cytokines are molecules that affect other cells and ligands (attaching molecules).

Cryoablation is tissue killing by freezing.

Dedifferentiated cells are cells further away in their appearance from normal cells: they are found in tumours considered to be aggressive.

Dendritic cells (DCs) are the most efficient antigen antigen-presenting cells found in epithelia and most other tissues.

Denosumab (Prolia, Xgeva) is a monoclonal antibody that targets RANKL, a cytokine essential for the formation, function, and survival of osteoclasts.

Detrusor is the name of bladder wall muscle which that normally contracts (only) during voiding.

DHEA dehydroepiandrosterone is an androgen produced by cells in the by the zona reticularis of the adrenal cortex.

Diapedesis is the passage of cells through the intact walls of the capillaries.

Dihydrotestosterone (DHT) which, together with testosterone (**T**), are the most common androgens. DHT, formed through the enzymic action of 5 alpha reductase, has a higher binding affinity for the androgen receptor (**AR**) than T.

DNA (deoxyribonucleic acid) is a molecule composed of two nucleic acid chains coiled around each other in a double helix.

DRE Digital rectal examination was previously referred to as a PR (per-rectal) examination.

EBR external beam radiotherapy.

ECOG Eastern Cooperative Oncology Group.

ED (erectile dysfunction) is difficulty in getting or maintaining an erection for satisfactory intercourse.

Endothelium is the term given to the lining cells of vessels.

Enzymes are biological catalysts.

EPIC is the acronym for the expanded prostate cancer index composite scores which is a validated, self-administered questionnaire used to assess the sexual function of men who had whole gland treatments such as prostatectomy and radiotherapy.

Epigenome/epigenetic region of a gene serves as a switch to activate or silence genes (via methylation and acetylation histone changes in particular): epigenetic processes instruct the cell to read specific "pages" of the DNA code.

Epithelial Mesenchymal transition (EMT) is the process by which cells change shape and become motile, EMT being active during development of the embryo. In prostate cancer, "EMT genes can be switched-on" with the prostatic adenocarcinomas cells assuming a spindle appearance so they resemble stromal or mesenchymal cells.

Epitope is the part of an antigen that is recognised by an antibody or T cell.

Erectile dysfunction is difficulty in achieving or keeping and erection so that sexual intercourse is possible.

Eugonadal means having normally functioning gonads in terms of androgen production.

FDG PET is fluorodeoxyglucose positron- emission tomography is imaging with radiolabelled fluorodeoxyglucose.

FDG PET CT involves the use of CT to construct images formed by FDG PET.

Fossa is an anatomical depression.

Grays (abbreviation gys) are the units used for radiation therapy with one gray defined as the absorption of one joule of radiation energy per kilogram of matter.

Gynaecomastia/Gynecomastia is breast tissue development in males ("man boobs").

High High-Grade Prostate Intraepithelial Neoplasia HGPIN (often referred to as PIN) is an appearance identified by light microscopy believed to be a precursor of prostate cancer, given the strong association between HGPIN and prostatic adenocarcinomas and that more than 80% of prostates with adenocarcinomas also contain HGPIN.

Hyperplasia is an increase in size due to an increase in the number of component cells.

Hypoxia is lacking oxygen.

International Index of Erectile Function-5 (IIEF-5) and IIEF-15 are validated, self-administered questionnaires used to assess the sexual function of men who had whole gland treatments such as prostatectomy and radiotherapy.

HIFU is the abbreviation for high high-intensity focussed ultrasound.

ICIs is the abbreviation for Immune Checkpoint Inhibitors.

Immune senescence is a gradual deterioration of the immune system with ageing characterised by shrinkage of the T cell repertoire and a propensity to undergo apoptosis. Amongst other changes, immunosenescent T lymphocytes are less able to remove their accumulated cellular debris.

Index lesion is the lesion suspected to have the highest level of cancer cancer-based on multiparametric MRI imaging of the prostate.

ISUP is an abbreviation for the International Society of Urological Pathologists.

LHRH/GnRH Luteinising Hormone Releasing Hormone or Gonadotrophin Hormone Releasing Hormone.

Ligands are binding or attaching molecules.

LUTS is the acronym for lower urinary tract symptoms, the lower urinary tract being the bladder and below: ureters and kidneys constitute the upper urinary tract.

Malignant tumours/tumors or cancers are tumours with propensity to invade surrounding tissues *and* to form secondary tumours in other sites in the body separated from the primarly lesion by normal tissue.

MET or Mesenchymal Epithelial Transition is the reverse of EMT in which cancer cells elongate and squeeze between endothelial cells in capillaries to move to future secondary sites.

Metabolism means the chemical processes required to maintain life or viability: from Greek metabolē "change" (from metabelian "to change").

Metastases (also referred to as secondaries) are cancers in another site remote and separated from the primary lesion by non-cancerous tissue.

microRNAs are 21–25 nucleotide, noncoding RNAs that can either negatively or positively regulate gene expression in a sequence-specific manner at a post-transcriptional level. Each miRNA sequence regulates the expression of multiple mRNA sequences, and one mRNA can be regulated by multiple miRNA sequences.

Microsatellite A set of short, repeated DNA sequences at a particular locus on a chromosome. Microsatellites are present in all people.

Microsatellite instability (MSI) Microsatellite instability is the condition of genetic hypermutability (predisposition to mutation) that results from impaired DNA mismatch repair.

Microtubules an important component of the cell cytoskeleton, a microscopic network of protein filaments and tubules in the cytoplasm of living cells, that gives them shape.

Mismatch repair (MMR) is the process in normal cells for correcting mistakes in DNA by correcting spontaneously occurring errors during DNA replication. Mismatch repair proteins, such as MSH2, MSH6, MLH1 and PMS2, form a complex that binds to the mismatched segments of DNA, excises the error, and inserts the correct sequence.

MoDCs is an abbreviation for monocyte monocyte-derived dendritic cells.

Morbidity means suffering from a disease or from medication.

Myokines are cytokines origination from muscles.

MSI is an abbreviation for microsatellite instability.

Neoadjuvant androgen deprivation therapy (Neoadjuvant ADT) in relation to radiotherapy is ADT commenced before and continued during radiotherapy.

Neovascularisation new vessel formation.

Neurendocrine dedifferentiation in prostate cancer is found almost exclusively in castrate resistant disease and is a term used to describe tumour cells that appear (and function) as neuroendocrine rather than prostatic cancer cells.

NICE is the acronym for the National Institute for Health and Clinical Excellence, an advisory body for England and Wales established to promote the highest quality of treatment as well as cost-effectiveness for the NHS.

NO nitric oxide.

Noncoding RNAs are functional RNA molecules transcribed from DNA but not translated into proteins. They function in protein production by regulating gene activity.

Oestrogens/estrogens steroid hormones which promote the development and maintenance of female characteristics of the body.

Organ tropism is the movement of cancer cells to preferred sites for metastases.

Osteoblasts bone bone-producing cells.

Osteoclasts bone bone-resorbing cells.

Osteolytic (adj) noun osteolysis the pathological destruction or disappearance of bone tissue.

Osteonecrosis of the Jaw (ONJ) is a condition in which dead bone is visible in the mouth.

Osteopaenia (osteopenia) a lesser degree of bone substance loss than osteoporosis.

Osteoporosis loss of bone substance.

Osteosclerotic (adj) increased bone density (on X-ray or radiograph).

Paget's Disease is a chronic condition affecting the elderly in which there is a rapid turnover of bone with deposition of poorly mineralised, weakened bone prone to deformity and fracture. Bones of the skull, spine, pelvis and femora are commonly sites of involvement.

Paraplegia is the condition in which there is paralysis of the lower body, including lower limbs, typically caused by spinal injury or disease.

PARP is the acronym for the enzyme poly adenosine polyphosphate ribose polymerase.

PC is an abbreviation used for Palliative Care.

PDE5 I is the abbreviation commonly used for Phosphodiesterase Type 5 Inhibitors Sildenafil (Viagra), Vardenafil (Levitra, Staxyn), Tadalafil (Adcirca, Cialis).

PDL-1 is the abbreviation for programmed cell death ligand – 1.

Perforin Perforins and granzymes are preformed mediators contained in lysosome-like granules of cytotoxic T lymphocytes (CTLs). CTLs that recognise target cells secrete a complex consisting of perforin, granzymes, and other proteins that enter the target cells by endocytosis. In the target cell cytoplasm, perforin facilitates the release of granzymes, proteases that cleave and activate caspases, which induce apoptosis. (Activated cytotoxic CD8+ T cells also kill cancer cells by secreting the cytokines Tumour Necrosis Factor- α (**TNF α**) and Interferon-γ (**IFN γ**) and by Fas ligand (**FasL**) on the T-cell surface binding and activating Fas receptor expressed on the target cells).

PET (positron emission tomography) scans an imaging technology in which substances containing positron-emitting isotopes are introduced into the body, allowing the precise location of physiological processes by detection of the gamma rays produced by the isotopes.

Phagocytosis is ingestion or breaking down and is the specialised process in which cells engulf relatively large, solid material.

Phenotype is the observable characteristic resulting from an interaction of genotype and environment. As such, the phenotype is the physical expression of DNA and is used loosely to mean appearance as opposed to genotype which refers to the DNA "make-up."

PIRADS an acronym for the prostate imaging-reporting and data system. It is an internationally accepted, structured reporting system for MRI of the prostate used for indicating the likelihood of prostate cancer and, if present, provides a rating of the cancer's likely clinical significance.

Pituitary gland The pituitary gland or hypophysis is referred to as the master gland of the endocrine system. Situated at the base of the brain immediately below the hypothalamus with which it is intimately related anatomically and functionally, the pituitary has anterior and posterior parts with distinctly different functions. The anterior pituitary produces hormones that travel to the thyroid, adrenals, and testicles (as well as other structures) which in turn produce thyroxine, cortisol, and testosterone, respectively, amongst other hormones. Regulation of these hormones is via feedback loops to the hypothalamus and anterior pituitary. A hormone produced by the hypothalamus and released by the posterior pituitary, antidiuretic hormone or ADH, is vital in water resorption by the collecting ducts in the kidneys.

Plasma cells Each plasma cell, which is derived from an antigen-stimulated B cell, secretes antibodies that recognise the same antigen that bound to the receptor of its B cell predecessor.

Promoter region of a gene is the region of DNA where transcription is initiated. Promoter regions are typically upstream of the sense or coding strand of the transcribed gene.

Prostatic Acid Phosphatase (PAcP) is an enzyme produced by the prostate. It was the most used prostate cancer tumour marker prior to PSA becoming available.

PSA (Prostate Specific Antigen) (*KLK3*) is a member of the kallikrein family of genes and proteins produced by both benign and malignant prostatic tissue. Although a poor indicator of the presence of prostate cancer in diagnosis, the PSA blood test is a sensitive indicator of persisting prostatic tissue following treatment for prostate cancer. A rising PSA post-treatment is presumptive evidence of persisting cancer progressing with the PSA doubling time (**PDAdt**) considered a better indicator of outcome and long-term survival than a single PSA level.

PSCA is the abbreviation for prostate stem cell antigen.

PSMA (Prostate Specific Membrane Antigen) is a cell surface protein highly expressed by a majority of prostate cancer cells. Although also present on normal cells at a number of sites throughout the body, PSMA is overexpressed in most (but not all) prostate cancer cells. PSMA is radiolabelled for imaging and therapeutic interventions.

PTSD Post-traumatic stress disorder.

Radical Prostatectomy is the operation in which the whole prostate is removed along with the seminal vesicles and the immediately adjacent part of the vas deferens.

Radiographs are X-rays.

Radiotherapy/radiation therapy is the treatment of disease by ionising radiation. Radiotherapy can be delivered from outside the body known as external beam radiation therapy/external beam radiotherapy (**EBRT/EBR**) or teletherapy, by radioactive implants

(**Brachytherapy**) or radiation combined with a biomarker such as PSMA. The combination of a biomarker for imaging combined simultaneously or sequentially with the same biomarker radiolabelled to treat a condition is called **Theranostics. Stereotactic radiotherapy** (**SRT**) involves giving the radiation from different angles around the body with the beams meeting at the tumour so that the target site receives a much higher dose than the tissues around it.

RCTs randomised controlled trials.

Rhabdosphincter consists of specialised skeletal muscle with an abundance of slow-twitch fibres at the bottom of the prostate and overlapping its lower part to a variable degree. It relaxes during voiding, just prior to bladder-neck opening, to enable urine to pass into the urethra below.

RNA (ribonucleic acid) is a copy of one strand of the DNA molecule.

Messenger RNA (mRNA) contains genetic information in the form of codons (a sequence of three nucleotides that together form a unit of genetic code) which is transcribed from DNA in the cell nucleus.

Transfer RNA (tRNA) transfers amino acids to ribosomes in the cytoplasm for synthesising proteins.

ROS is the abbreviation for reactive oxygen species which are oxygen-containing radicals formed from oxygen.

Sarcomas are cancers that arise from stromal or connective tissue components of tissue cells as opposed to adenocarcinomas which are derived from epithelial (surface lining) cells.

Sarcopaenia (sarcopenia) is loss of skeletal muscle mass.

SEER (Surveillance, epidemiology, and end results) registry contains cancer incidence and survival data from population-based registries from approximately one-third of the US population.

Skeletal Related Events are unwanted changes related to bone secondaries and/or cancer treatment effects on bone.

Sphincter muscle surrounding and keeping a tube, in this context the urethra, closed at rest and relaxing immediately prior to and during voiding to allow passage of urine.

STHLM3 (Stockholm 3 test) is a test for prostate cancer reported to be superior to PSA. It includes five plasma protein markers (total PSA, free PSA, free hK2, PSP94, GDF-15), more than 100 genetic markers, and clinical data such as age, previous biopsy results, and family history to determine the risk of aggressive prostate cancer (defined as Gleason Score ≥ 7).

Stress urinary incontinence (also called stress incontinence) is unintentional loss of urine with physical movement or activity.

Technetium-99m is a nuclear isomer of technetium-99. Used as a radioactive tracer, it emits gamma rays of about the same wavelength emitted by conventional X-ray diagnostic equipment. The half-life is fractionally over 6 hours. Technetium-99m is attached to bisphonphonates for use in radio-isotope bone scans which, after injection, concentrate in areas of increased bone activity.

Testosterone (**T**) is a male hormone (androgen) produced predominantly by Leydig cells in the testis.

Thromboembolic a process in which a thrombus (blood clot) forms and moves to another site.

TNM/AJCC is the abbreviation for the Union for International Cancer Control Tumour Node Metastases/American Joint Committee on Cancer and refers to the staging system agreed upon and used world-wide in prostate (and other) cancers. The preceding letters are "c" (for clinical) and "p" (for pathological), respectively.

Tumour/tumor is a mass of abnormal tissue that forms in the body. Tumours are either benign or malignant (cancer).

Thymus gland T cells (more correctly called T lymphocytes) are so named as they are derived from the **thymus gland**, situated in the neck and upper thorax: the thymus decreases in size, in a process known as involution, at about the time of puberty. Mature T cells are found in the blood, where they constitute 60-70% of lymphocytes, and in T-cell zones of peripheral lymphoid organs.

T lymphocytes (also called T cells)

Helper T cells or lymphocytes stimulate B lymphocytes to proliferate and maturate to antibody-producing plasma cells in addition to activating other leukocytes.

Cytotoxic T cells CD8+ cytotoxic T cells recognise cell-bound antigens only in association with class I MHC (major histocompatibility complex) molecules and kill cells through perforin and granzyme, Fas – FasL interaction, IFγ, and TNFα (as above).

NK (Natural Killer) cells recognise and destroy severely stressed or abnormal cells, such as cancer cells. NK cells make up approximately 5%–10% of peripheral blood lymphocytes.

Regulatory T cells (Tregs) are a T-lymphocyte subtype that suppresses activation, proliferation, and cytokine production by CD4+ and CD8+ cells.

Trabeculae are partitions formed by bands of connective tissue which, in bone, consist of denser areas of osseus tissue.

Transition Zone (TZ) is the part of the prostate in which BPH develops. Prostate cancer can also develop in the TZ but, when arising from this zone, the cancers are usually clinically insignificant and an incidental finding by the pathologist.

Transurethral Resection of the Prostate (TURP) or "rebore" operation involves using an instrument (called a resectoscope) inserted through the urethra to resect tissue from the inside so that the prostatic urethral channel is widened.

Tricarboxylic acid (TCA) or citric acid cycle, eponymously known as the Krebs cycle is the three-stage process by which living cells break down organic fuel molecules in the presence of oxygen to provide essential energy.

TRUS is the abbreviation for transrectal ultrasound (imaging) of the prostate.

References

[1] Prostate cancer (canceraustralia.gov.au), https://www.canceraustralia.gov.au/cancer-types/prostate-cancer/overview.

[2] Prostate cancer in Australia statistics, Cancer Australia, https://www.canceraustralia.gov.au/cancer-types/prostate-cancer/statistics.

[3] https://www.abs.gov.au/statistics/health/causes-death/causes-death-australia/2020.

[4] O. Bratt, What the Urologist should know about hereditary predisposition to prostate cancer, BJU Int. 99 (4) (2007) 743–747.

[5] I. Agalliu, E. Karlins, E.M. Kwon, et al., Rare germline mutations in the BRCA2 gene are associated with early-onset prostate cancer, Br. J. Cancer 97 (6) (2007) 826–831.

[6] V.L. Patel, E.L. Busch, T.M. Friebel, et al., Association of genomic domains in BRCA1 and BRCA2 with prostate cancer risk and aggressiveness, Cancer Res. 80 (3) (2020) 624–638.

[7] T. Nyberg, D. Frost, D. Barrowdale, et al., Prostate cancer risks for male BRCA1 and BRCA2 mutation carriers: a prospective cohort study, Eur. Urol. 77 (1) (2020) 24–35.

[8] L.G. Gordon, H.W. Tuffaha, R. James, et al., Estimating the healthcare costs of treating prostate cancer in Australia: A Markov modelling analysis, Urol. Oncol. 36 (3) (2018) 91.e7-91.e15, doi:10.1016/j.urolonc.2017.10.024.

[9] T.R. McClintock, E.B. Cone, M. Marchese, et al., Prostate cancer management costs vary by disease stage at presentation, Prostate Cancer Prostatic Dis. 23 (2020) 564–566.

[10] M.J. Roberts, P. Teloken, S.K. Chambers, et al. Prostate Cancer Detection. In: Endotext [Internet]. South Dartmouth, MA: MDText.com, Inc.; 2000–2018 Jun 11. De Groot L.J., Chrousos G., Dungan K., et al., Eds. https://www.endotext.org/section/male/.

[11] T. Callender, M. Emberton, S. Morris, et al., Benefit, harm, and cost-effectiveness associated with magnetic resonance imaging before biopsy in age-based and risk-stratified screening for prostate cancer, JAMA Netw. Open 4 (3) (2021) e2037657, doi:10.1001/jamanetworkopen.2020.37657.

[12] F.H. Drost, D. Osses, D. Nieboer, et al., Prostate MRI, with or without MRI-targeted biopsy, and systematic biopsy for detecting prostate cancer, Cochrane Database Syst. Rev. 2019 (4) (2019) CD012663.

[13] D.C. Wedge, G. Gundem, T. Mitchell, et al., Sequencing of prostate cancers identifies new cancer genes, routes of progression and drug targets, Nat. Genet. 50 (5) (2018 May) 682–692.

[14] W.A. Sakr, D.J. Grignon, J.D. Crissman, et al., High grade prostatic intraepithelial neoplasia (HGPIN) and prostatic adenocarcinoma between the ages of 20-69: an autopsy study of 249 cases, In Vivo 8 (3) (1994) 439–443.

[15] A.B. Weiner, R.S. Matulewicz, S.E. Eggener, E.M. Schaeffer, Increasing incidence of metastatic prostate cancer in the United States (2004-2013), Prostate Cancer Prostatic Dis. 19 (4) (2016) 395–397.

[16] S.P. Kelly, W.F. Anderson, P.S. Rosenberg, M.B. Cook, Past, current, and future incidence rates and burden of metastatic prostate cancer in the United States, Eur. Urol. Focus 4 (1) (2018) 121–127.

[17] A.O. Elmehrath, A.M. Afifi, M.J. Al-Husseini, et al., Causes of Death Among Patients With Metastatic Prostate cancer in the US from 200 to 2016, JAMA Netw. Open 4 (8) (2021) e2119568, doi:10.1001/jamanetworkopen.2021.19568.

[18] S.W.D. Merriel, T. Martins, S.E.R. Bailey, Exploring the causes of death among patients with metastatic prostate cancer—a changing landscape, JAMA Netw. Open 4 (8) (2021) e2120889.

[19] J. Walz, A. Gallina, P. Perotte, et al., Clinicians are poor raters of life-expectancy before radical prostatectomy or definitive radiotherapy for localized prostate cancer, BJU Int. 100 (6) (2007) 1254–1258.

[20] R. Singh, T.S. O'Brien, Comorbidity assessment in localized prostate cancer: a review of currently available techniques, Eur. Urol. 46 (1) (2004) 28–41.

[21] M.S. Litwin, S. Greenfield, E.P. Elkin, et al., Assessment of prognosis with the total illness burden index for prostate cancer: aiding clinicians in treatment choice, Cancer 109 (9) (2007) 1777–1783.

[22] A.B. Araujo, T.G. Travison, P. Ganz, et al., Erectile dysfunction and mortality, J. Sex Med. 6 (2009) 2445–2454.

[23] K.K. Chew, N. Gibson, F. Sanfilippo, et al., Cardiovascular mortality in men with erectile dysfunction: increased risk but not inevitable, J. Sex Med. 8 (6) (2011) 1761–1771.

[24] H. von Staden, The discovery of the body: human dissection and its cultural contexts in ancient Greece, Yale J. Biol. Med. 65 (1992) 223–241.

[25] D.B. Moffat, Development of the urogenital system in the male. In: Scientific Foundations of Urology, second edition, G.D. Chisholm, D. Innes Williams Heinemann, Eds. (1982) p 352.

[26] J.E. McNeal, E.A. Redwine, F.S. Freiha, T.A. Stamey, Zonal distribution of prostatic adenocarcinoma. Correlation with histologic pattern and direction of spread, Am. J. Surg. Pathol. 12 (12) (1988) 897–906.

[27] H. Bonkhoff, N. Wernert, G. Dhom, K. Remberger, Basement membranes in fetal, adult normal, hyperplastic and neoplastic human prostate, Virchows Arch. A Pathol. Anat. Histopathol. 418 (1991) 375–381.

[28] M.A. Kosir, W. Wang, K.L. Zukowski, G. Tromp, Degradation of basement membrane by prostate tumor heparanase, J. Surg. Res. 81 (1999) 42–47.

[29] U.H. Stenman, J. Leinonen, W.M. Zhang, P. Finne, Prostate-specific antigen, Semin. Cancer Biol. 9 (2) (1999) 83–93.

[30] J.A. Clements, T. Merritt, K. Devoss, et al., Inactive free:total prostate specific antigen ratios in ejaculate from men with suspected and known prostate cancer, compared with young control men, BJUI Int. 86 (4) (2000) 453–458.

[31] M.J. Lynch, J. Masters, J.P. Pryor, et al., Ultra high field NMR spectroscopic studies on human seminal fluid, seminal vesicle and prostatic secretions, J. Pharm. Biomed. Anal. 12 (1994) 5–19.

[32] G.P. Liney, L.W. Turnbull, M. Lowry, et al., In vivo quantification of citrate concentration and water T2 relaxation time of the pathologic prostate gland using 1H MRS and MRI, Magn. Reson. Imaging 15 (1997) 1177–1186.

[33] M.J. Roberts, H.J. Schirra, M.F. Lavin, R.A. Gardiner, Metabolomics: a novel approach to early and noninvasive prostate cancer detection, Korean J. Urol. 52 (2) (2011) 79–89.

[34] B. Debrabant, M. Soerensen, F. Flachsbart, et al., Human longevity and variation in DNA damage response and repair: study of the contribution of sub-processes using competitive gene-set analysis, Eur. J. Hum. Genet. 22 (9) (2014) 1131–1136.

[35] H. Boughey, M. Jurga, S.F. El-Khamisy, DNA homeostasis and senescence: Lessons from the naked mole rat, Int. J. Mol. Sci. 22 (11) (2021) 6011, doi:10.3390/ijms22116011.

[36] Z. Shang, J. Yu, L. Sun, et al., LncRNA PCAT1 activates AKT and NF-κB signaling in castration-resistant prostate cancer by regulating the PHLPP/FKBP51/IKKα complex, Nucleic. Acids. Res. 47 (8) (2019) 4211–4225.

[37] A.A.I. Sina, L.G. Carrascosa, Z. Kuang, et al., Epigenetically reprogrammed methylation landscape drives the DNA self-assembly and serves as a universal cancer biomarker, Nat. Commun. 9 (1) (2018) 4915.

[38] R. Thomas, M. Holm, The benefits of exercise after cancer-an international review of the clinical and microbiological benefits, Br. J. Med. Pract. 1 (2014) 2–9.

[39] L. Puca, P.J. Vlachostergios, H. Beltran, Neurendocrine Differentiation in prostate cancer: emerging biology, models and therapies, Cold Spring Harb. Perspect. Med. 9 (2) (2019) a030593.

[40] G. Kroemer, G. Mariño, B. Levinne, Autophagy and the integrated stress response, Mol. Cell 40 (2) (2010) 280–293.

[41] Y. Li, L.X. Pang, Z. Cui, et al., Tumor-derived autophagosome vaccine: mechanism of cross-presentation and therapeutic efficacy, Clin. Cancer Res. 17 (22) (2011) 7047–7057.

[42] S. Fais, M. Overholtzer, Cell-in-cell phenomena, cannibalism and autophagy: is there a relationship? Cell Death. Dis. 9 (2) (2018) 95, doi:10.1038/s41419-017-0111-7.

[43] T. Jamaspishvili, D.M. Berman, A.E. Ross, et al., Clinical implications of PTEN loss in prostate cancer, Nat. Rev. Urol. 15 (4) (2018) 222–234.

[44] S. Ben-Salem, V.B. Venkadakrishnan, H.V. Heemers, Novel insights in cell cycle dysregulation during prostate cancer progression, Endocr. Relat. Cancer 28 (6) (2021) R141–R155, doi:10.1530/ERC-20-0517.

[45] D.I. Quinn, P.D. Stricker, J.G. Kench, et al., p53 nuclear accumulation as an early indicator of lethal prostate cancer, Br. J. Cancer 121 (7) (2019) 578–583.

[46] P.J. Campbell, G. Getz, J.O. Korbel, et al., Pan-cancer analysis of whole genomes, Nature 578 (7793) (2020) 82–93.

[47] J. Göhring, F. Kellner, L. Schrangl, et al., Temporal analysis of T-cell receptor-imposed forces via quantitative single molecule FRET measurements, Nat. Commun. 12 (1) (2021) 2502, doi:10.1038/s41467-021-22775-z.

[48] A. Lanzavecchia, Antigen-specific interaction between T and B cells, Nature 17:314 (6011) (1985) 537–539, doi:10.1038/314537a0.

[49] A. Trefzer, P. Kadam, S.-H. Wang, et al., Dynamic adoption of anergy by antigen-exhausted CD4+ T cells, Cell Rep. 34 (6) (2021), doi:10.1016/j.celrep.2021.108748.

[50] M.S. Abdel-Hakeem, S. Manne, J.-C. Beltra, et al., Epigenetic scarring of exhausted T cells hinders memory differentiation upon eliminating chronic antigenic stimulation, Nat. Immunol. 22 (8) (2021) 1008–1019, doi:10.1038/s41590-021-00975-5.

[51] A. Cachot, M. Bilous, Y.-C. Liu, et al., Tumor-specific cytolytic CD4 T cells mediate immunity against human cancer, Sci. Adv. 7 (9) (2021) eabe3348, doi:10.1126/sciadv.abe3348.

[52] S. Sheppard, E.K. Santosa, C.M. Lau, et al., Lactate dehydrogenase A-dependent aerobic glycolysis promotes natural killer cell anti-viral and anti-tumor function, Cell Rep. 35 (9) (2021) 109210, doi:10.1016/j.celrep.2021.109210.

[53] E. Sikora, Activation-induced and damage-induced cell death in aging human T cells, Mech. Ageing Dev. 151 (2015) 85–92.

[54] G. Alsaleh, I. Panse, L. Swadling, et al., Autophagy in T cells from aged donors is maintained by spermidine, and correlates with function and vaccine responses, Elife 9 (2020) e57950, doi:10.7554/eLife.57950.

[55] S. Arora, P.J. Thompson, Y. Wang, et al., Invariant natural killer T cells coordinate removal of senescent cells, Med. (NY) 2 (8) (2021) 938–950, doi:10.1016/j.medj.2021.04.014.

[56] R. Dittrich, J. Kurth, E.A. Decelle, et al., Assessing prostate cancer growth with citrate measured by intact tissue proton magnetic resonance spectroscopy, Prostate Cancer Prostatic Dis. 15 (3) (2012) 278–282.

[57] M. Clyne, Prostate cancer: biopsy citrate concentration could predict prostate cancer growth rate, Nat. Rev. Urol. 9 (3) (2012) 123, doi:10.1038/nurol,2012.24.

[58] M.L. Lynch, J.K. Nicholson, Proton MRS of human prostatic fluid: correlations between citrate, spermine and myo-inositol levels and changes with disease, Prostate 30 (4) (1997) 248–255.

[59] M.J. Roberts, H.J. Schirra, M.F. Lavin, R.A. Gardiner, Metabolomics: a novel approach to early and non-invasive prostate cancer detection: invited review, Korean J. Urol. 52 (2) (2011) 79–89.

[60] L.C. Costello, R.B. Franklin, Prostatic fluid electrolyte composition for the screening of prostate cancer: a potential solution to a major problem, Prostate Cancer Prostatic Dis. 12 (1) (2009) 17–24.

[61] A.F. Hansen, E. Sandsmark, M.B. Rye, et al., Presence of TMPRSS2-ERG is associated with alterations of the metabolic profile in human prostate cancer, Oncotarget 7 (27) (2016) 42071–42085, doi:10.18632/oncotarget.9817.

[62] J.E. Cooper, I. Farid, The role of citric acid in the physiology of the prostate: lactic/citrate ratios in benign and malignant prostatic homogenates as an index of prostatic malignancy, J. Urol. 92 (1964) 533–536.

[63] F.K. Habib, M.K. Mason, P.H. Smith, S.R. Stitch, Cancer of the prostate: early diagnosis by zinc and hormone analysis, Br. J. Cancer 39 (1979) 700–704.

[64] M. Damaghi, J. West, M. Robertson-Tessi, et al., The harsh microenvironment in early breast cancer selects for a Warburg phenotype, Proc. Natl. Acad. Sci. 118 (3) (2021) e2011342118, doi:10.1073/pnas.2011342118.

[65] F.J. Naser, M.M. Jackstadt, R. Fowle-Grider, et al., Isotope tracing in adult zebrafish reveals alanine cycling between melanoma and liver, Cell Metab. 33 (7) (2021) 1493–1504.e5, doi:10.1016/j.cmet.2021.04.014.

[66] H. Alshaker, K. Sacco, A. Alfraidi, et al., Leptin signalling, obesity and prostate cancer: molecular and clinical perspective on the old dilemma, Oncotarget 6 (34) (2015) 35556–35563.

[67] R. Shi, H.J. Berke., H. Yu, Insulin-like growth factor-I and prostate cancer: a meta-analysis, Br. J. Cancer 85 (7) (2001) 991–996.

[68] R. Virtakoivu, J.H. Rannikko, M. Viitala, et al., Systemic blockade of Clever-1 elicits lymphocyte activation alongside checkpoint molecule downregulation in patients with solid tumors: Results from a phase I/II clinical trial, Clin. Cancer Res. 27 (15) (2021) 4205–4220, doi:10.1158/1078-0432.CCR-20-4862.

[69] Z. Bian, L. Shi, K. Kidder, et al., Intratumoural SIRPα-deficient macrophages activate tumor antigen-specific cytotoxic T cells under radiotherapy, Nat. Commun. 12 (1) (2021) 3229, doi:10.1038/s41467-021-23442-z.

[70] T. Le Bourgeois, L. Strauss, H.-I. Aksoylar, et al., Targeting T cell metabolism for improvement of cancer immunotherapy, Front. Oncol. 8 (2018) 237, doi:10.3389/fonc.2018.00237.

[71] E. Giampazolias, O. Schulz, K.H.J. Lim, et al., Secreted gelsolin inhibits DNGR-1-dependent cross-presentation and cancer immunity, Cell 184 (15) (2021) 4016–4031.e22, doi:10.1016/j.cell.2021.05.021.

[72] A.E. Ringel, J.M. Drijvers, G.J. Baker, et al., Obesity shapes metabolism in the tumor microenvironment to suppress anti-tumor immunity, Cell 183 (7) (2020) 1848–1866.e26, doi:10.1016/j.cell.2020.11.009.

[73] X. Michelet, L. Dyck, A. Hogan, et al., Metabolic reprogramming of natural killer cells in obesity limits antitumor responses, Nat. Immunol. (2018) 1330–1340, doi:10.1038/s41590-018-0251-7.

[74] Y. Zhao, X. Fu, J.I. Lopez, et al., Selection of metastasis competent subclones in the tumour interior, Nat. Ecol. Evol. 5 (7) (2021) 1033–1045, doi:10.1038/s41559-021-01456-6.

[75] K. Gravdal, O.J. Halvorsen, S.A. Haukaas, L.A. Akslen, A switch from E-cadherin to N-cadherin expression indicates epithelial to mesenchymal transition and is of strong and independent importance for the progress of prostate cancer, Clin. Cancer Res. 13 (23) (2007) 7003–7011.

[76] K. Polyak, R.A. Weinberg, Transitions between epithelial mesenchymal states: acquisition of malignant and stem cell traits, Nat. Rev. Cancer 9 (2009) 265–273.

[77] G. Larocque, D.J. Moore, M. Sittewelle, et al., Intracellular nanovesicles mediate $\alpha 5\beta 1$ integrin trafficking during cell migration, J. Cell Biol. 220 (10) (2021) e202009028, doi:10.1083/jcb.202009028.

[78] N. Aceto, A. Bardia, D.T. Miyamoto, et al., Circulating tumor cell clusters are oligoclonal precursors of breast cancer metastasis, Cell 158 (5) (2014) 1110–1122.

[79] F. Chemi, S. Mohan, T. Guevara, et al., Early dissemination of circulating tumor cells: biological and clinical insights, Front. Oncol 11 (2021) 672195, doi:10.3389/fonc.2021.672195.

[80] R.N. Kaplan, R.D. Riba, S. Zacharoulis, et al., VEGFR1-positive haematopoietic bone marrow progenitors initiate the pre-metastatic niche, Nature 438 (7069) (2005) 820–827, doi:10.1038/nature04186.

[81] A. Recasens, L. Munoz, Targeting cancer cell dormancy, Trends Pharmacol. Sci. 40 (2) (2019) 128–141.

[82] F.M. Vallette, C. Olivier, F. Lézot, et al., Dormant, quiescent, tolerant and persister cells: four synonyms for the same target in cancer, Biochem. Pharmacol. 162 (2019) 169–176, doi:10.1016/j.bcp.2018.11.004.

[83] A.L. Correia, J.C. Guimaraes, P. Auf der Maur, et al., Hepatic stellate cells suppress NK cell-sustained breast cancer dormancy, Nature (2021) 1–6, doi:10.1038/s41586-021-03614-z.

[84] B. Bao, A. Ahmad, D. Kong, et al., Hypoxia induced aggressiveness of prostate cancer cells is linked with deregulated expression of VEGF, IL-6 and miRNAs that are attenuated by CDF, PLoS One 7 (8) (2012) e43726, doi:10.1371/journal.pone.0043726.

[85] A. Patrikidou, Y. Loriot, J.-C. Eymard, et al., Who dies from prostate cancer? Prostate Cancer Prostatic Dis. 17 (4) (2014) 348–352, doi:10.1038/pcan.2014.35.

[86] K. Fall, F. Fang, L.A. Mucci, et al., Immediate risk for cardiovascular events and suicide following a prostate cancer diagnosis: prospective cohort study, PLoS Med. 6 (12) (2009) e1000197.

[87] J. Dunn, S.K. Chambers, Feelings, and feelings, and feelings. Let me try thinking instead': Screening for distress and referral to psychosocial care for men with prostate cancer, Eur. J. Cancer Care. 28 (5) (2019) e13163, doi:10.1111/ecc.13163.

[88] I.J. Korfage, M.L. Essink-Bot, G.J. Borsboom, et al., Five-year follow-up of health-related quality of life after primary treatment of localized prostate cancer, Int. J. Cancer 116 (2) (2005) 291–296, doi:10.1002/ijc.21043.

[89] G. Ilie, J. White, R. Mason, et al., Current Mental Distress Among Men With a History of Radical Prostatectomy and Related Adverse Correlates, Am. J. Men' Health 14 (5) (2020) 1557988320957535, doi:10.1177/1557988320957535.

[90] P.J. Han, C. Gutheil, R.N. Hutchinson, J.A. LaChance, Cause or effect? The role of prognostic uncertainty in the fear of cancer recurrence, Front. Psychol. 11 (2021) 626038, doi:10.3389/fpsyg.2020.626038.

[91] F. Crawford-Williams, S. March, B.C. Goodwin, et al., Interventions for prostate cancer survivorship: a systematic review of reviews, Psychooncology 27 (10) (2018) 2339–2348.

[92] J. Dunn, A. Green, N. Ralph, et al., Prostate cancer survivorship essentials framework: guidelines for practitioners, BJU Int. 128 (Suppl 3) (2021) 18–29, doi:10.1111/bju.15159.

[93] D.A. Galvão, R.U. Newton, S.K. Chambers, et al., Psychological distress in men with prostate cancer undertaking androgen deprivation therapy: modifying effects of exercise from a year-long randomised controlled trial, Prostate Cancer Prostatic Dis. 24 (3) (2021) 758–766.

[94] P. Cormie, E.M. Zopf, Z. Zhang, K. Schmitz, The impact of exercise on cancer mortality, recurrence, and treatment-related adverse effects, Epidemiol. Rev. 39 (1) (2017) 71–92.

[95] D.R. Taafe, R. Newton, N. Spry, et al., Effects of different exercise modalities on fatigue in prostate cancer patients undergoing androgen deprivation therapy: a year-long randomised controlled trial, Eur. Urol. 72 (2) (2017) 293–299.

[96] D.G. Bostwick, A. Pacelli, M. Blute, et al., Prostate specific membrane antigen expression in prostatic intraepithelial neoplasia and adenocarcinoma: a study of 184 cases, Cancer 82 (11) (1998) 2256–2261.

[97] S. Mannweiler, P. Amersdorfer, S. Trajanoski, et al., Heterogeneity of prostate-specific membrane antigen (PSMA) expression in prostate carcinoma with distant metastasis, Pathol. Oncol. Res. 15 (2) (2009) 167–172.

[98] H.J. Ananias, M.C. van den Heuvel, W. Helfrich, et al., Expression of the gastrin-releasing peptide receptor, the prostate stem cell antigen and the prostate-specific membrane antigen in lymph node and bone metastases of prostate cancer, Prostate 69 (10) (2009) 1101–1108.

[99] S.S. Chang, V.E. Reuter, W.D. Heston, et al., Five different anti-prostate-specific membrane antigen (PSMA) antibodies confirm PSMA expression in tumor-associated neovasculature, Cancer Res. 59 (13) (1999) 3192–3198.

[100] R.L. Sokoloff, K.C. Norton, C.L. Gasior, et al., A dual-monoclonal sandwich assay for prostate-specific membrane antigen: levels in tissues, seminal fluid and urine, Prostate 43 (2) (2000) 150–157.

[101] N.H. Bander, Technology insight: monoclonal antibody imaging of prostate cancer, Nat. Clin. Pract. Urol. 3 (4) (2006) 216–225.

[102] D.M. Schuster, B. Savir-Baruch, P.T. Nieh, et al., Detection of recurrent prostate carcinoma with anti-1-amino-3-18F-fluorocyclobutane-1-carboxylic Acid PET/CT and 111In–capromab pendetide SPECT/CT, Radiology 259 (3) (2011) 852–861, doi:10.1148/radiol.11102023.

[103] P. Donato, M.J. Roberts, A. Morton, et al., Improved specificity with 68Ga PSMA PET/CT to detect clinically significant lesions "invisible" on multiparametric MRI of the prostate: a single institution comparative analysis with radical prostatectomy histology, Eur. J. Nucl. Med. Mol. Imaging 6 (1) (2019) 20–30.

[104] J.W. Yaxley, S. Raveenthiran, F.-X. Nouhaud, et al., Outcomes of primary lymph node staging of intermediate and high risk prostate cancer with 68 Ga-PSMA positron emission tomography/computerized tomography compared to histological correlation of pelvic lymph node pathology, J. Urol. 201 (4) (2019) 815–820.

[105] M.S. Hofman, N. Lawrentschuk, R.J. Francis, et al., Prostate-specific membrane antigen PET-CT in patients with high-risk prostate cancer before curative-intent surgery or

radiotherapy (proPSMA): a prospective, randomised, multicentre study, Lancet 395 (10231) (2020) 1208–1216.

[106] A. Franklin, W.J. Yaxley, S. Raveenthiran, et al., Histological comparison between predictive value of preoperative 3-T multiparametric MRI and 68 Ga-PSMA PET/CT scan for pathological outcomes at radical prostatectomy and pelvic lymph node dissection for prostate cancer, BJU Int. 127 (1) (2021) 71–79.

[107] D. Meijer, P.J. van Leeuwen, M.J. Roberts, et al., External validation and addition of prostate-specific membrane antigen positron emission tomography to the most frequently used nomograms for the prediction of pelvic lymph-node metastases: an international multicenter study, Eur. Urol. 80 (2) (2021) 234–242.

[108] M. Farsad, R. Schiavina, P. Castellucci, et al., Detection and localization of prostate cancer: correlation of (11)C-choline PET/CT With histopathologic step-section analysis, J. Nucl. Med. 46 (10) (2005) 1642–1649.

[109] C. Nanni, R. Schiavina, E. Brunocilla, et al., 18f-Fluciclovine PET/CT for the detection of prostate cancer relapse: a comparison to 11C-Choline PET/CT, Clin. Nucl. Med. 40 (8) (2015) e386–e391.

[110] S.C. Kamran, J.A. Efstathiou, Current state of personalized genitourinary cancer radiotherapy in the era of precision medicine, Front. Oncol. 11 (2021) 675311, doi:10.3389/fonc.2021.675311.

[111] R. Aggarwal, S.C. Behr, P.L. Paris, et al. Real-Time Transferrin-Based PET Detects MYC-Positive Prostate Cancer. Mol. Cancer Res. 15 (9) (2017) 1221–1229.

[112] A.B. Jani, E. Schreibmann, S. Goyal, et al., 18 F-fluciclovine-PET/CT imaging versus conventional imaging alone to guide postprostatectomy salvage radiotherapy for prostate cancer (EMPIRE-1): a single centre, open-label, phase 2/3 randomised controlled trial, Lancet 397 (10288) (2021) 1895–1904.

[113] R.E. de Feria Cardet, M.S. Hofman, T. Segard, et al., Is prostate-specific membrane antigen positron emission tomography/computed tomography imaging cost-effective in prostate cancer: An analysis informed by the proPSMA trial, Eur. Urol. 79 (3) (2021) 413–418, doi:10.1016/j.eurorp.2020.11.043.

[114] X.R.S. Ong, D. Bagguley, J.W. Yaxley, Understanding the diagnosis of prostate cancer, Med. J. Aust. 213 (9) (2020) 424–429.

[115] H.W. Børg, The Evidence-based transformation of american medicine, J. American Physicians Surgeons 21 (3) (2016) 70–73.

[116] H. Chen, M. Twiddy, L. Jones, M.J. Johnson, The unique information and communication needs of men affected by prostate cancer: A qualitative study of men's experience, Eur. J. Cancer Care 30 (6) (2021) e13503, doi:10.1111/ecc.13503.

[117] B. Brown, J. Young, D.P. Smith, et al., A multidisciplinary team-oriented intervention to increase guideline recommended care for high-risk prostate cancer: A stepped-wedge cluster randomised implementation trial, Implement Sci. 13 (2018) 43, doi:10.1186/s13012-018-0733-x.

[118] J.I. Epstein, P.C. Walsh, M. Carmichael, et al., Pathologic and clinical findings to predict tumor extent of nonpalpable (stage T1c) prostate cancer, JAMA 271 (1994) 368–374.

[119] P.R. Carroll, J.K. Parsons, G. Andriole, et al., Prostate cancer early detection, version 1.2014. Featured updates to the NCCN guidelines, J. Natl. Compr. Canc. Netw. 12 (9) (2014) 1211–1219, doi:10.6004/jccn.2014.0120.

[120] I. Thompson, J.B. Thrasher, G. Aus, et al., Guideline for the management of clinically localized prostate cancer: 2007 update, J. Urol. 177 (6) (2007) 2106–2131, doi:10.1016/j.juro.2007.03.003.

[121] M.G. Sanda, J.A. Cadeddu, E. Kirkby, et al., Clinically localized prostate cancer: AUA/ASTRO/SUO guideline. Part I: risk stratification, shared decision making, and care options, J. Urol. 199 (3) (2018) 683–690, doi:10.1016/j.juro.2017.11.095.

[122] R.T. Dess, K. Suresh, M.J. Zelefsky, et al., Development and validation of a clinical prognostic stage group system for nonmetastatic prostate cancer using disease-specific mortality results from the international staging collaboration for cancer of the prostate, JAMA Oncol. 6 (12) (2020) 1912–1920.

[123] Y.A. Nyame, W. Underwood III, Development and validation of a clinical prognostic stage group system for nonmetastatic prostate cancer using disease-specific mortality results from the international staging collaboration for cancer of the prostate, JAMA Oncol. (2020), doi:10.1001/jamaoncol.2020.4916.

[124] M.R. Cooperberg, S.J. Freedland, D.J. Pasta, et al., Multiinstitutional validation of the UCSF cancer of the prostate risk assessment for prediction of recurrence after radical prostatectomy, Cancer. 107 (10) (2006) 2384–2391.

[125] J. Cuzick, D.M. Berney, G. Fisher, et al., Prognostic value of a cell cycle progression signature for prostate cancer death in a conservatively managed needle biopsy cohort, Br. J. Cancer 106 (6) (2012) 1095–1099.

[126] J. Cuzick, G.P. Swanson, G. Fisher, et al., Prognostic value of an RNA expression signature derived From cell cycle proliferation genes in patients with prostate cancer: a retrospective study, Lancet Oncol. 12 (3) (2011) 245–255.

[127] M.R. Cooperberg, J.P. Simko, J.E. Cowan, et al., Validation of a cell-cycle progression gene panel to improve risk stratification in a contemporary prostatectomy cohort, J. Clin. Oncol. 31 (11) (2013) 1428–1434.

[128] NICE Advice - prolaris gene expression assay for assessing long-term risk of prostate cancer progression: © NICE (2016) prolaris gene expression assay for assessing long-term risk of prostate cancer progression, BJU Int. 122 (2) (2018) 173–180.

[129] P. Blume-Jensen, D.M. Berman, D.L. Rimm, et al., Development and clinical validation of an in situ biopsy-based multimarker assay for risk stratification in prostate cancer, Clin. Cancer Res. 21 (11) (2015) 2591–2600.

[130] E.A. Klein, M.R. Cooperberg, C. Magi-Galluzzi, et al., A 17-Gene assay to predict prostate cancer aggressiveness in the context of Gleason grade heterogeneity, tumor multifocality, and biopsy undersampling, Eur. Urol. 66 (3) (2014) 550–560.

[131] J. Cullen, I.L. Rosner, T.C. Brand, et al., A biopsy-based 17-gene genomic prostate score predicts recurrence after radical prostatectomy and adverse surgical pathology in a racially diverse population of men with clinically low- and intermediate-risk prostate cancer, Eur. Urol. 68 (1) (2015) 123–131.

[132] D.E. Spratt, K. Yousefi, S. Deheshi, et al., Individual patient-level meta-analysis of the performance of the decipher genomic classifier in high-risk men after prostatectomy to predict development of metastatic disease, J. Clin. Oncol. 35 (18) (2017) 1991–1998.

[133] E.A. Klein, D. Richards, A. Cohn, et al., Clinical validation of a targeted methylation-based multi-cancer early detection test using an independent validation set, Annals Oncol. 32 (9) (2021) 1167–1177, doi:10.1016/j.annonc.2021.05.806.

[134] R.J. Bryant, J. Oxley, G.J. Young, et al., ProtecT Study Group. The ProtecT trial: analysis of the patient cohort, baseline risk stratification and disease progression, BJU Int. 125 (4) (2020) 506–514, doi:10.1111/bju.14987.

[135] F.C. Hamdy, J.L. Donovan, J.A. Lane, et al., 10-year outcomes after monitoring, surgery, or radiotherapy for localized prostate cancer, N. Engl. J. Med. 375 (15) (2016) 1415–1424, doi:10.1056/NEJMoa1606220.

[136] D.E. Neal, C. Metcalfe, J.L. Donovan, et al., Ten-year mortality, disease progression, and treatment-related side effects in men with localised prostate cancer from the protect randomised controlled trial according to treatment received, Eur. Urol. 77 (3) (2020) 320–330, doi:10.1016/j.eururo.2019.10.030.

[137] A. Aggarwal, J. Nossiter, M. Parry, et al., Public reporting of outcomes in radiation oncology: the National Prostate Cancer Audit, Lancet Oncol. 22 (5) (2021) e207–e215, doi:10.1016/S1470-2045(20)30558-1.

[138] S. Raveenthiran, J. Yaxley, T. Gianduzzo, et al., The use of 68Ga PET PSMA to determine patterns of disease for biochemically recurrence prostate cancer following radiotherapy, Prostate Cancer Prostatic Dis. 22 (3) (2019) 385–390, doi:10.1038/s41391-019-0163-0.

[139] A.M. el-Shirbiny, Prostatic specific antigen, Adv. Clin. Chem. 31 (1994) 99–133.

[140] F. Rae, B. Bulmer, D. Nicol, J. Clements, The human tissue kallikreins (KLKs 1-3) and a novel KLK1 mRNA transcript are expressed in a renal cell carcinoma cDNA library, Immunopharmacology 45 (1-3) (1999) 83–88.

[141] N.J. Freeman, C. Doolittle, Elevated prostate markers in metastatic small cell carcinoma of unknown primary, Cancer 68 (5) (1991) 1118–1120.

[142] A.W. Partin, C.R. Pound, J.Q. Clemens, et al., Serum PSA after anatomic radical prostatectomy. The Johns Hopkins experience after 10 years, Urol. Clin. North Am. 20 (4) (1993) 713–725.

[143] C.R. Pound, A.W. Partin, J.I. Epstein, P.C. Walsh, Prostate-specific antigen after anatomic radical retropubic prostatectomy: patterns of recurrence and cancer control, Urol. Clin. North Am. 24 (2) (1997) 395–406.

[144] A.J. Stephenson, M.W. Kattan, J.A. Eastham, et al., Defining biochemical recurrence of prostate cancer after radical prostatectomy: a proposal for a standardized definition, J. Clin. Oncol. 24 (24) (2006) 3973–3978.

[145] R.A. Gardiner, J. Yaxley, G. Coughlin, et al., A randomised trial of Robotic and open prostatectomy in men with localised prostate cancer, BMC Cancer 12 (2012) 189, doi:10.1186/1471-2407-12-189.

[146] S.J. Freedland, M.E. Sutter, F. Dorey, W.J. Aronson, Defining the ideal cutpoint for determining PSA recurrence after radical prostatectomy. Prostate-specific antigen, Urology 61 (2) (2003) 365–369.

[147] A.V. D'Amico, J.W. Moul, P.R. Carroll, et al., Surrogate end point for prostate cancer-specific mortality after radical prostatectomy or radiation therapy, J. Natl. Cancer Inst. 95 (18) (2003) 1376–1383.

[148] P.B. Romesser, X. Pei, W. Shi, et al., Prostate-specific antigen (PSA) bounce after dose-escalated external beam radiation therapy is an independent predictor of PSA recurrence, metastasis, and survival in prostate adenocarcinoma patients, Int. J. Radiat. Oncol. Biol. Phys. 100 (1) (2018) 59–67.

[149] F.A. Critz, W.H. Williams, J.B. Benton, et al., Prostate specific antigen bounce after radioactive seed implantation followed by external beam radiation for prostate cancer, J. Urol. 163 (4) (2000) 1085–1089.

[150] Y. Yamamoto, C.P. Offord, G. Kimura, et al., Tumour and immune cell dynamics explain the PSA bounce after prostate cancer brachytherapy, Br. J. Cancer 115 (2) (2016) 195–202.

[151] M. Roach 3rd, G. Hanks, H. Thames, et al., Defining biochemical failure following radiotherapy with or without hormonal therapy in men with clinically localized prostate cancer: recommendations of the RTOG-ASTRO Phoenix Consensus Conference, Int. J. Radiat. Oncol. Biol. Phys. 65 (4) (2006) 965–974.

[152] M.C. Abramowitz, T. Li, M.K. Buyyounouski, et al., The Phoenix definition of biochemical failure predicts for overall survival in patients with prostate cancer, Cancer 112 (1) (2008) 55–60.
[153] J.E. Bekelman, R.B. Rumble, R.C. Chen, et al., Clinically localized prostate cancer: ASCO clinical practice guideline endorsement of an american urological association/american society for radiation oncology/society of urologic oncology guideline, J. Clin. Oncol. 36 (32) (2018) 3251–3258.
[154] J.L. Mohler, E.S. Antonarakis, A.J. Armstrong, et al., Prostate cancer, Version 2.2019, NCCN clinical practice guidelines in oncology, J. Natl. Compr. Canc. Netw. 17 (5) (2019) 479–505, doi:10.6004/jnccn.2019.0023.
[155] N. Mottet, J. Bellmunt, M. Bolla, et al., EAU-ESTRO-SIOG guidelines on prostate cancer. Part 1: screening, diagnosis, and local treatment with curative intent, Eur. Urol. 71 (4) (2017) 618–629, doi:10.1016/j.eururo.2016.08.003.
[156] M.G. Sanda, J.A. Cadeddu, E. Kirkby, et al., Clinically Localized prostate cancer: AUA/ASTRO/SUO Guideline. Part II: Recommended approaches and details of specific care options, J. Urol. 199 (4) (2018) 990–997, doi:10.1016/j.juro.2018.01.002.
[157] A. Finelli, M. Komisarenko, L.J. Martin, et al., Long-term use of 5-alpha-reductase inhibitors is safe and effective in men on active surveillance for prostate cancer, Prostate Cancer Prostatic Dis. 24 (1) (2021) 69–76, doi:10.1038/s41391-020-0218-2.
[158] S.W.D. Merriel, L. Hetherington, A. Seggie, et al., Best practice in active surveillance for men with prostate cancer: a Prostate Cancer UK consensus statement, BJU Int. 124 (1) (2019) 47–54, doi:10.1111/bju.14707.
[159] M.A. Evans, J.L. Millar, A. Earnest, et al., Active surveillance of men with low risk prostate cancer: evidence from the Prostate Cancer Outcomes Registry–Victoria, Med. J. Aust. 208 (10) (2018) 439–443.
[160] L. Klotz, Contemporary approach to active surveillance for favorable risk prostate cancer, Asian J. Urol. 6 (2) (2019) 146–152, doi:10.1016/j.ajur.2018.12.003.
[161] A.A. Kalapara, J.F.M. Verbeek, D. Neiboer, et al., Adherence to active surveillance protocols for low-risk prostate cancer: results of the movember foundation's global action plan prostate cancer active surveillance initiative, Eur. Urol. Oncol. 3 (1) (2020) 80–91, doi:10.1016/j.euo.2019.08.014.
[162] A. Heidenreich, P.J. Bastian, J. Bellmunt, et al., EAU Guidelines on prostate cancer. Part 1: screemomg, diagnosis and local treatment with curative intent – Update 2013, Eur. Urol. 65 (2014) 124–137. https://uroweb.org/wp-content/uploads/09_Prostate_Cancer_LR.pdf.
[163] L.P. Bokhorst, A.R. Alberts, A. Rannikko, et al., Compliance rates with the prostate cancer research international active surveillance (PRIAS) protocol and disease reclassification in noncompliers, Eur. Urol. 68 (5) (2015) 814–821.
[164] V. da Silva, I. Cagiannos, L.T. Lavallée, et al., An assessment of prostate cancer research international: active surveillance (PRIAS) criteria for active surveillance of clinically low-risk prostate cancer patients, Can. Urol. Assoc. J. 1 (8) (2017) 238–243.
[165] J.S. Detsky, A.F. Ghiam, A. Mamedov, et al., Impact of biopsy compliance on outcomes for patients on active surveillance for prostate cancer, J. Urol. 204 (5) (2020) 934–940.
[166] D.-W. Kang, A.S. Fairey, N.G. Boulé, et al., Effects of exercise on cardiorespiratory fitness and biochemical progression in men with localized prostate cancer under active surveillance: the ERASE randomized clinical trial, JAMA Oncol. 7 (10) (2021) 1487–1495, doi:10.1001/jamaoncol.2021.3067.
[167] K. Koo, J. Wang, R. Richards, et al., Design and clinical verification of surface enhanced raman spectroscopy diagnostic technology for individual cancer risk prediction, ACS Nano 12 (8) (2018) 8362–8371, doi:10.1021/acsnano.8b03698.

[168] J. Yaxley, J. Dagher, B. Delahunt, et al., Reconsidering the role of pelvic lymph node dissection in an era of improving radiological staging techniques, World J. Urol. 36 (1) (2018) 15–20.

[169] K.A. Touijer, D.D. Sjoberg, N. Benfante, et al., Limited versus extended pelvic lymph node dissection for prostate cancer: a randomized clinical trial, Eur. Urol. Oncol. 4 (4) (2021) 532–539, doi:10.1016/j.euo.2021.03.006.

[170] J.W. Yaxley, G.D. Coughlin, S.K. Chambers, et al., Robot-assisted laparoscopic prostatectomy versus open radical retropubic prostatectomy: early outcomes from s randomised controlled phase 3 study, Lancet 388 (10049) (2016) 1057–1066.

[171] G.D. Coughlin, J.W. Yaxley, S.K. Chambers, et al., Robot-assisted laparoscopic prostatectomy versus open radical retropubic prostatectomy: 24-month outcomes from a randomised controlledstudy, Lancet Oncol. 19 (8) (2018) 1051–1060.

[172] V. Sharma, R.J. Karnes, Much ado about robotic versus open radical prostatectomy, Lancet Oncol. 19 (8) (2018) 1003–1004.

[173] V. Ficarra, G. Novara, R.C. Rosen, et al., Systematic review and meta-analysis of studies reporting urinary continence recovery after robot-assisted radical prostatectomy, Eur. Urol. 62 (2012) 405–417.

[174] P.S. Moran, M. O'Neill, C. Teljeur, et al., Robot-assisted radical prostatectomy compared with open and laparoscopic approaches: a systematic review and meta-analysis, Int. J. Urol. 20 (3) (2013) 312–321.

[175] A. Tamhankar, N. Spencer, A. Hampson, et al., Real-time assessment of learning curve for robot-assisted laparoscopic prostatectomy, Ann. R. Coll. Surg. Engl. 102 (9) (2020) 717–725.

[176] E. Checcucci, A. Veccia, C. Fiori, et al., Retzius-sparing robot-assisted radical prostatectomy vs the standard approach: a systematic review and analysis of comparative outcomes, BJU Int. 125 (1) (2020) 8–16, doi:10.1111/bju.14887.

[177] P. Umari, C. Eden, D. Cahill, et al., Retzius-sparing versus standard robot-assisted radical prostatectomy: a comparative prospective study of nearly 500 patients, J. Urol. 205 (3) (2021) 780–790, doi:10.1097/JU.0000000000001435.

[178] J. Egan, S. Marhamati, F.L.F. Carvalho, et al., Retzius-sparing robot-assisted radical prostatectomy leads to durable improvement in urinary function and quality of life versus standard robot-assisted radical prostatectomy without compromise on oncologic efficacy: single-surgeon series and step-by-step guide, Eur. Urol. 79 (6) (2021) 839–857.

[179] R. Propert, D. Cahill, C. Eden, Advances in robotic prostatectomy. Urology News, Prostate Cancer Focus. Ed. Propert R (2021) 13–15.

[180] N. Schifano, P. Capogrosso, M. Tutolo, et al., How to Prevent and Manage Post-Prostatectomy Incontinence: A Review, World J. Mens. Health 39 (4) (2021) 581–597.

[181] S. Herschorn, J. Thuroff, H. Bruschini, et al., Surgical treatment of urinary incontinence in men, in: Incontinence Proceedings of the 3rd International Consultation on Incontinence, Vol 2, Health Publication Ltd., 2005, pp. 1241–1296.

[182] D. Perucchini, J. DeLancey, J. Ashton-Miller, et al., Age effects on urethral striated muscle. 1. Changes in number and diameter of striated muscle fibers in the ventral urethra, Am. J. Obstet. Gynecol. 186 (3) (2002) 351–355.

[183] M. Baunacke, M.-L. Schmidt, C. Thomas, et al., Long-term functional outcomes after robotic vs. retropubic radical prostatectomy in routine care: A 6-year follow-up of a large German health services research study, World J. Urol. 38 (2020) 1701–1709.

[184] E. Haglind, S. Carlsson, J. Stranne, et al., Urinary incontinence and erectile dysfunction after robotic versus open radical prostatectomy: a prospective, controlled, nonrandomised trial, Eur. Urol. 68 (2015) 216–225.

[185] M.C. Moschovas, S. Bhat, F.F. Onol, et al., Modified apical dissection and lateral prostatic fascia preservation improves early postoperative functional recovery in robotic-assisted laparoscopic radical prostatectomy: results from a propensity score-matched analysis, Eur. Urol. 78 (6) (2020) 875–884.
[186] K.V. Carlson, V.W. Nitti, Prevention and management of incontinence following radical prostatectomy, Urol. Clin. North Am. 28 (3) (2001) 595–612.
[187] D. Barski, H. Gerullis, T. Otto, Review of surgical implant procedures for male incontinence after radical prostatectomy according to IDEAL framework, Updates Surg. 69 (3) (2017) 327–338.
[188] J.I. Chang, V. Lam, M.I. Patel, Preoperative pelvic floor muscle exercise and postprostatectomy incontinence: a systematic review and meta-analysis, Eur. Urol. 69 (3) (2016) 460–467.
[189] F. Abdollah, M. Sun, N. Suardi, et al., Prediction of functional outcomes after nerve-sparing radical prostatectomy: results of conditional survival analyses, Eur. Urol. 62 (1) (2012) 42–52.
[190] A.J. Vickers, M. Kent, J. Mulhall, J. Sanghu, Counseling the post radical prostatectomy patients about functional recovery: high predictiveness of current status, Urology 84 (1) (2014) 158–163.
[191] A. Aydın Sayılan, A. Özbaş, The effect of pelvic floor muscle training on incontinence problems after radical prostatectomy, Am. J. Mens. Health 12 (4) (2018) 1007–1015.
[192] J.E. Milios, T.R. Ackland, D.J. Green, Pelvic floor muscle training in radical prostatectomy: a randomized controlled trial of the impacts on pelvic floor muscle function and urinary incontinence, BMC Urol. 19 (1) (2019) 116.
[193] K. Matsushita, B.I. Chughtai, A.C. Maschino, R.K. Lee, J.S. Sandhu, International variation in artificial urinary sphincter use, Urol. 80 (3) (2012) 667–672.
[194] L. Guorong, C. Lambert, A. Gentil-Perret, et al., Molecular and cytometric analysis of renal cell carcinoma cells. Concepts, techniques and prospects, Prog. Urol. 13 (1) (2003) 1–13.
[195] M. Nyberg, D.D. Sjoberg, S.V. Carlsson, et al., Surgeon heterogeneity significantly affects functional and oncological outcomes after radical prostatectomy in the Swedish LAPPRO trial, BJU Int. 127 (3) (2021) 361–368.
[196] C. Vale, D. Fisher, A. Kneebone, et al., Adjuvant or early salvage radiotherapy for the treatment of localised and locally advanced prostate cancer: a prospectively planned systematic review and meta-analysis of aggregate data, Lancet 396 (10260) (2020) 1422–1431.
[197] C.C. Parker, N.W. Clarke, A.D. Cook, et al., Timing of radiotherapy after radical prostatectomy (RADICALS-RT): a randomised, controlled phase 3 trial, Lancet 396 (10260) (2020) 1413–1421.
[198] N. Sundahl, A. Tree, C. Parker, The emerging role of local therapy in metastatic prostate cancer, Curr. Oncol. Rep. 22 (1) (2020) 2.
[199] C. Carrie, N. Magné, P. Burban-Provost, et al., Short-term androgen deprivation therapy combined with radiotherapy as salvage treatment after radical prostatectomy for prostate cancer (GETUG-AFU 16): a 112-month follow-up of a phase 3, randomised trial, Lancet Oncol. 20 (12) (2019) 1740–1749.
[200] P.S. Heathcote, P.N. Mactaggart, R.J. Boston, et al., Health-related quality of life in Australian men remaining disease-free after radical prostatectomy, Med. J. Aust. 168 (1998) 483–486.
[201] G. Fiard, D. Kelly, T. Yap, M. Emberton, Detailing sexual outcomes after treatment of localised prostate cancer with focal therapy using various energy sources: protocol for a mixed-methods study, BMJ Open 10 (12) (2020) e045500.

[202] A.E. Isgoren, T.R. Saitz, E.C. Serefoglu, Erectile function outcomes after robot-assisted radical prostatectomy: is it superior to open retropubic or laparoscopic approach? Sex Med. Rev. 2 (1) (2014) 10–23.

[203] M. Sia, G. Rodrigues, C. Menard, et al., Treatment-related toxicity and symptom-related bother following postoperative radiotherapy for prostate cancer, Can. Urol. Assoc. J. 4 (2) (2010 Apr) 105–111.

[204] M. Doumas, S. Douma, The effect of antihypertensive drugs on erectile function: a proposed management algorithm, J. Clin. Hypertens (Greenwuch) 8 (5) (2006) 359–364.

[205] A.J. Bella, G.B. Brock, Intracavernous pharmacotherapy for erectile dysfunction, Endocrine 23 (2-3) (2004) 149–155.

[206] C. Liu, D.S. Lopez, M. Chen, R. Wang, Penile rehabilitation therapy following radical prostatectomy: a meta-analysis, J. Sex Med. 14 (12) (2017) 1496–1503.

[207] A. Salonia, G. Asaikan, J. Buvat, et al., Sexual rehabilitation after treatment for prostate cancer—Part 1: recommendations From the Fourth International Consultation for Sexual Medicine (ICSM 2015), J. Sex. Med. 14 (2017) 297–315.

[208] S. Vacas, D.J. Cole, M. Cannesson, Cognitive decline associated with anesthesia and surgery in older patients, JAMA (2021), doi:10.1001/jama.2021.4773.

[209] E. Mahanna-Gabrielli, K.J. Schenning, L.I. Eriksson, et al., State of the clinical science of perioperative brain health: report from the American Society of Anesthesiologists Brain Health Initiative Summit 2018, Br. J. Anaesth. 123 (4) (2019) 464–478.

[210] K. Chan, S. Morris, S. Hughes, S. Harris, Advances in radiotherapy techniques for prostate cancer: shorter, faster, sharper and safer treatments, Urology News Prostate cancer focus. Ed. Propert R (2021) 13–15.

[211] T.M. Pawlik, K. Keyomarsi, Role of cell cycle in mediating sensitivity to radiotherapy, Int. J. Radiat. Oncol. Biol. Phys. 59 (4) (2004) 928–942.

[212] https://ieer.org/resource/classroom/measuring-radiation-terminology/#:~:text=Some%20units%20used%20in%20measuring%20ionizing%20radiation%20and,%3D%20100%20rad%20%203%20more%20rows%20.

[213] P. Fransson, P. Nilsson, A. Gunnlaugsson, et al., Ultra-hypofractionated versus conventionally fractionated radiotherapy for prostate cancer (HYPO-RT-PC): Patient-reported quality-of-life outcomes of a randomised, controlled, non-inferiority, phase 3 trial, Lancet Oncol. 22 (2) (2021) 235–245.

[214] A. Nabid, N. Carrier, E. Vigneault, et al., Androgen deprivation therapy and radiotherapy in intermediate-risk prostate cancer: a randomised phase III trial, Eur. J. Cancer 143 (2021) 64–74.

[215] S. Supiot, C. Rousseau, M. Dore, et al., Evaluation of tumor hypoxia prior to radiotherapy in intermediate-risk prostate cancer using 18F-fluoromisonidazole PET/CT: a pilot study, Oncotarget 9 (11) (2018) 0005–10015.

[216] P. Wardman, Nitroimidazoles as hypoxic radiosensitizers and hypoxia probes: misonidazole, myths and mistakes, Br. J. Radiol. 92(1093) (2019) 20170915.

[217] L.H. Gray, A.D. Conger, M. Ebert, et al., The concentration of oxygen dissolved in tissues at the time of irradiation as a factor in radiotherapy, Br. J. Radiol. 26 (312) (1953) 638–648.

[218] H.B. Ragnum, L. Vlatkovic, A.K. Lie, et al., The tumour hypoxia marker pimonidazole reflects a transcriptional programme associated with aggressive prostate cancer, Br. J. Cancer 112 (2) (2015) 382–390.

[219] N.J. Rupp, P.J. Schüffler, Q. Zhong, et al., Oxygen supply maps for hypoxic microenvironment visualization in prostate cancer, J. Pathol. Inform. 7 (2016) 3, doi:10.4103/2153-3539.175376.

[220] O. Schumacher, D.A. Galvão, D.R. Taaffe, et al., Exercise modulation of tumour perfusion and hypoxia to improve radiotherapy response in prostate cancer, Prostate Cancer Prostatic Dis. 24 (1) (2021) 1–14.

[221] D.J. McCullough, J.N. Stabley, D.W. Siemann, B.J. Behnke, Modulation of blood flow hypoxia and vascular function in orthotopic prostate tumors during exercise, J. Natl. Cancer Inst. 106 (4) (2014) dju036, doi:10.1093/jnci/dju036.

[222] R.M. Johnke, J.M. Edwards, C.J. Kovacs, et al., Response of T lymphocyte populations in prostate cancer patients undergoing radiotherapy: influence of neoajuvant total androgen suppression, Anticancer Res. 25 (4) (2005) 3159–3166.

[223] S.P. Keam, H. Halse, T. Nguyen, et al., High dose-rate brachytherapy of localized prostate cancer converts tumors from cold to hot, J. Immunother. Cancer 8 (1) (2020) e000792.

[224] Y. Ma, R. Conforti, L. Aymeric, et al., How to improve the immunogenicity of chemotherapy and radiotherapy, Cancer Metastasis Rev. 30 (1) (2011) 71–82.

[225] R.G. Bristow, B. Alexander, M. Baumann, et al., Combining precision radiotherapy with molecular targeting and immunomodulatory agents: a guideline by the American Society for Radiation Oncology, Lancet Oncol. 195 (5) (2018) e240–e251, doi:10.1016/S1470-2045(18)30096-2.

[226] J.M. Taylor, V.E. Chen, R.C. Miller, B.A. Greenberger, The impact of prostate cancer treatment on quality of life: a narrative review with a focus on randomized data, Res. Reports Urol. 12 (2020) 533–546.

[227] D.E. Spratt, S. Malone, S. Roy, et al., Prostate radiotherapy with adjuvant androgen deprivation therapy (ADT) improves metastasis-free survival compared to neoadjuvant ADT: an individual patient meta-analysis, J. Clin. Oncol. 39 (2) (2021) 136–144.

[228] M. Bolla, T.M. de Reijke, G. Van Tienhoven, et al., Duration of androgen suppression in the treatment of prostate cancer, N. Engl. J. Med. 360 (24) (2009) 2516–2527.

[229] D. Joseph, J.W. Denham, A. Steigler, et al., Radiation dose escalation or longer androgen suppression to prevent distant progression in men with locally advanced prostate cancer: 10-Year data from the TROG 03.04 RADAR trial, Int. J. Radiat. Oncol. Biol. Phys. 106 (4) (2020) 693–702.

[230] K. Dacal, S.M. Sereika, S.L. Greenspan, Quality of life in prostate cancer patients taking androgen deprivation therapy: quality of life in men with prostate cancer, J. Am. Geriatr. Soc. 54 (1) (2006) 85–90.

[231] A.S. Cheung, C. de Rooy, R. Hoermann, et al., Quality of life decrements in men with prostate cancer undergoing androgen deprivation therapy, Clin. Endocrinol. (Oxf) 86 (3) (2017) 388–394.

[232] B. Nascimento, E.P. Miranda, L.C. Jenkins, et al., Testosterone recovery profiles after cessation of androgen deprivation therapy for prostate cancer, J. Sex Med. 16 (6) (2019) 872–879, doi:10.1016/j.jsxm.2019.03.273.

[233] A.V. D'Amico, W. Xie, E. McMahon, et al., Radiation and androgen deprivation therapy with or without docetaxel in the management of nonmetastatic unfavorable-risk prostate cancer: a prospective randomized trial, J. Clin. Oncol. 39 (26) (2021) 2938–2947, doi:10.1200/JCO.21.00596.

[234] S.G. Williams, J.M.G. Taylor, N. Liu, et al., Use of individual fraction size data from 3756 patients to directly determine the alpha/beta ratio of prostate cancer, Int. J. Radiat. Oncol. Biol. Phys. 68 (1) (2007) 24–33.

[235] H.A. Gay, H.J. Barthold, E. O'Meara, et al., Pelvic normal tissue contouring guidelines for radiation therapy: a Radiation Therapy Oncology Group consensus panel atlas, Int. J. Radiat. Oncol. 83 (3) (2012) e353–e362, doi:10.1016/j.ijrobp.2012.01.023.

[236] L.J. Bell, T. Eade, A. Kneebone, et al., Initial experience with intra-fraction motion monitoring using Calypso guided volumetric modulated arc therapy for definitive prostate cancer treatment, J. Med. Radiat. Sci. 64 (1) (2017) 25–34.

[237] P.J. Keal, J.A. Ng, P. Juneja, et al., Real-time 3D image guidance using a standard LINAC: measured motion, accuracy, and precision of the first prospective clinical trial of kilovoltage intrafraction monitoring-guided gating for prostate cancer radiation therapy, Int. J. Radiat. Oncol. Biol. Phys. 94 (5) (2016) 1015–1021.

[238] M.S. Peach, T.N. Showalter, N. Ohri, Systematic review of the relationship between acute and late gastrointestinal toxicity after radiotherapy for prostate cancer, Prostate Cancer 2015 (2015) 624736.

[239] J. Tulk, J.A. Rash, J. Thoms, et al., Androgen deprivation therapy and radiation for prostate cancer—cognitive impairment, sleep, symptom burden: a prospective study, BMJ Support Palliat Care (2021), doi:10.1136/bmjspcare-2021-003098.

[240] D.E. Guy, H. Chen, R. Boldt, et al., Characterizing surgical and radiotherapy outcomes in non-metastatic high-risk prostate cancer: a systematic review and meta-analysis, Cureus 13 (8) (2021) e17400, doi:10.7759/cureus.17400.

[241] A.C. Lo, W.J. Morris, T. Pickles, et al., Patterns of recurrence after low dose rate prostate brachytherapy: A population based on study of 2223 consecutive low and intermediate risk patients, Int. J. Radiat. Oncol. Biol. Phys. 91 (4) (2015) 745–751.

[242] P. Zhang, B. Qian, J. Shi, Y. Xia, Radical prostatectomy versus brachytherapy for clinically localized prostate cancer on oncological and functional outcomes: a meta-analysis, Transl. Androl. Urol. 9 (2) (2020) 332–343.

[243] G.D. Smith, T. Pickles, J. Crook, et al., Brachytherapy improves biochemical failure-free survival in low- and intermediate-risk prostate cancer compared with conventionally fractionated external beam radiation therapy: a propensity score matched analysis, Int. J. Radiat. Oncol. Biol. Phys. 91 (3) (2015) 505–516.

[244] S. Rodda, S. Tyldesley, W.J. Morris, et al., ASCENDE-RT: an analysis of treatment-related morbidity for a randomized trial comparing a low-dose-rate brachytherapy boost with a dose-escalated external beam boost for high- and intermediate-risk prostate cancer, Int. J. Radiat. Oncol. Biol. Phys. 98 (2) (2017) 286–295.

[245] D. Pasalic, D.A. Barocas, L.-C. Huang, et al., Five year outcomes from a prospective comparative effectiveness study evaluating external beam radiotherapy with or without low-dose-rate brachytherapy boost for prostate cancer, Cancer 127 (11) (2021) 1912–1925, doi:10.1002/cnr.3388.

[246] C. Buron, B. Le Vu, J.M. Cosset, et al., Brachytherapy versus prostatectomy and localised prostate cancer: Results of a French multicentre prospective medico-economic study, Int. J. Radiat. Oncol. Biol. Phys. 67 (3) (2007) 812–822.

[247] M. Keyes, S. Miller, T. Pickles, et al., Late urinary side effects 10 years after low-dose-rate prostate brachytherapy: population-based results from a multi-physician practice treating with a standardised protocol and uniform dosimetric goals, Int. J. Radiat. Oncol. Biol. Phys. 90 (3) (2014) 570–578.

[248] J.W. Yaxley, K. Lah, J.P. Yaxley, et al., Long-term outcomes of high-dose-rate brachytherapy for intermediate- and high-risk prostate cancer with a median follow-up of 10 years, BJU Int. 120 (1) (2017) 56–60, doi:10.1111/bju.13659.

[249] R.C. Borges, R.R. Tourinho-Barbosa, S. Glina, et al., Impact of focal versus whole gland ablation for prostate cancer on sexual function and urinary continence, J. Urol. 205 (1) (2021) 129–136.

[250] NICE (National Institute for Health and Care Excellence): Focal therapy using cryoablation and focal therapy using high-intensity focused ultrasound for localised prostate cancer https://www.nice.org.uk/guidance/published?type=apg,csg,cg,cov,mpg,ph,sg,sc,dg,hst,ipg,mtg,qs,ta&title=prostate%20cancer.
[251] A.-R. Azzouzi, S. Vincendeau, E. Barret, et al., Padeliporfin vascular-targeted photodynamic therapy versus active surveillance in men with low-risk prostate cancer (CLIN1001 PCM301): an open-label, phase 3, randomised controlled trial, Lancet Oncol. 18 (2) (2017) 181–191.
[252] Y. He, P. Tan, M. He, et al., The primary treatment of prostate cancer with high-intensity focused ultrasound: a systematic review and meta-analysis, Medicine (Baltimore). 99 (41) (2020) e22610.
[253] V. Berge, L. Dickinson, N. McCartan, et al., Morbidity associated with primary high intensity focused ultrasound and redo high intensity focused ultrasound for localized prostate cancer, J. Urol. 191 (6) (2014) 1764–1769.
[254] S. Guillaumier, M. Peters, M. Arya, et al., A multicentre study of 5-year outcomes following focal therapy in treating clinically significant nonmetastatic prostate cancer, Eur. Urol. 74 (4) (2018) 422–429.
[255] W. Van den Bos, M.J. Scheltema, A.R. Siriwardana, et al., Focal irreversible electroporation as primary treatment for localized prostate cancer, BJU Int. 121 (5) (2018) 716–724.
[256] D. Reddy, H.U. Ahmed, Focal therapy trials, Urology News 25 (2020) 23–24.
[257] G.P. Swanson, M.A. Hussey, C.M. Tangen, et al., SWOG 8794. Predominant treatment failure in postprostatectomy patients is local: analysis of patterns of treatment failure in SWOG 8794, J. Clin. Oncol. 25 (16) (2007) 2225–2229.
[258] T.D. Richardson, K.J. Wojno, L.W. Liang, et al., Half-life determination of serum free prostate-specific antigen following radical prostatectomy, Urology 48 (6A Suppl) (1996) 40–44.
[259] A.B. Weiner, E.V. Li, A.S. Desai, et al., Cause of death during prostate cancer survivorship: a contemporary, US population–based analysis, Cancer 127 (16) (2021) 2895–2904.
[260] A.O. Elmehrath, A.M. Afifi, M.J. Al-Husseini, et al., Causes of death among patients with metastatic prostate cancer in the US from 2000 to 2016, JAMA Netw. Open 4 (8) (2021) e2119568, doi:10.1001/jamanetworkopen.2021.19568.
[261] L.H.M. Rydzewska, S. Burdett, C.L. Vale, et al., Adding abiraterone to androgen deprivation therapy in men with metastatic hormone-sensitive prostate cancer: a systematic review and meta-analysis, Eur. J. Cancer 84 (2017) 88–101.
[262] E. Efstathiou Abstract LBA5_PR. Presented at ESMO Annual Meeting September 2021.
[263] R. Aggarwal, N.R. Rydzewski, L. Zhang, et al., Prognosis associated with luminal and basal subtypes of metastatic prostate cancer, JAMA Oncol. 7 (11) (2021) 1644–1652, doi:10.1001/jamaoncol.2021.3987.
[264] A.A. Hamid, H.-C. Huang, V. Wang, et al., Transcriptional profiling of primary prostate tumor in metastatic hormone-sensitive prostate cancer and association with clinical outcomes: correlative analysis of the E3805 CHAARTED trial, Ann. Oncol. 32 (9) (2021) 1157–1166.
[265] N.C. Bennett, R.A. Gardiner, J.D. Hooper, et al., Molecular cell biology of androgen receptor signalling, Int. J. Biochem. Cell Biol. 42 (6) (2010) 813–827.
[266] E. Nevedomskaya, S.J. Baumgart, B. Haendler, Recent advances in prostate cancer treatment and drug discovery, Int. J. Mol. Sci. 19 (5) (2018) 1359, doi:10.3390/ijms19051359.
[267] M. Singh, R. Jha, J. Melamed, et al., Stromal androgen receptor in prostate development and cancer, Am. J. Pathol. 184 (10) (2014) 2598–2607.

[268] S.M. Henshall, D.I. Quinn, C.S. Lee, et al., Altered expression of androgen receptor in the malignant epithelium and adjacent stroma is associated with early relapse in prostate cancer, Cancer Res. 61 (2) (2001) 423–427.

[269] C. Ricciardelli, C.S. Choong, G. Buchanan, et al., Androgen receptor levels in prostate cancer epithelial and peritumoral stromal cells identify non-organ confined disease, Prostate 63 (1) (2005) 19–28.

[270] S.M. Harman, E.J. Metter, J.D. Tobin, et al., Longitudinal effects of aging on serum total and free testosterone levels in healthy men. Baltimore Longitudinal Study of Aging, J. Clin. Endocrinol. Metab. 86 (2) (2001) 724–731.

[271] A.B. Araujo, G.R. Esche, V. Kupelian, et al., Prevalence of symptomatic androgen deficiency in men, J. Clin. Endocrinol. Metab. 92 (2007) 4241–4247.

[272] S.-M. Ho, M.-T. Lee, H.-M. Lam, Y.-K. Leung, Estrogens and prostate cancer: etiology, mediators, prevention and management, Endocrinol. Metab Clin North Am. 40 (3) (2011) 591–614.

[273] C. Huggins, C.V. Hodges, Studies on prostatic cancer. The effect of castration, of estrogen, and of androgen injection on serum phosphates in metastatic carcinoma of the prostate, Cancer Res. 1 (1) (1941) 293.

[274] C. Huggins, R.E. Stevens, C.V. Hodges, Studies of prostate cancer: II. The effects of castration on advanced carcinoma of the prostate gland, Arch. Surg. 43 (1941) 209–228.

[275] N. Shore, F. Saad, M.S. Cookson, et al., Oral Relugolix for androgen-deprivation therapy in advanced prostate cancer, N. Engl. J. Med. 382 (23) (2020) 2187–2196.

[276] P.W. Kantoff, S. Halabi, M. Conaway, et al., Hydrocortisone with or without mitoxantrone in men with hormone-refractory prostate cancer: results of the cancer and leukemia group B 9182 study, J. Clin. Oncol. 17 (8) (1999) 2506–2513.

[277] J.A. Kish, R. Bukkapatnam, F. Palazzo, The treatment challenge of hormone-refractory prostate cancer, Cancer Control 8 (6) (2001) 487–495.

[278] D. Schultheiss, R. Badalyan, A. Pilatz, et al., Androgen and estrogen receptors in the human corpus cavernosum penis: immunohistochemical and cell culture results, World J. Urol. 21 (5) (2003) 320–324.

[279] D.M. Janoff, C. Peterson, S. Mongoue-Tchokote, et al., Clinical outcomes of androgen deprivation as the sole therapy for localized and locally advanced prostate cancer, BJU Int. 96 (4) (2005) 503–507.

[280] P. Kantoff, Recent progress in management of advanced prostate cancer, Oncology (Williston Park) 19 (5) (2005) 631–636.

[281] C. Carson 3rd, R. Rittmaster, The role of dihydrotestosterone in benign prostatic hyperplasia, Urology 61 (4 Suppl) (2003) 2–7.

[282] B. Chatterjee, The role of the androgen receptor in the development of prostatic hyperplasia and prostate cancer, Mol. Cell. Biochem. 253 (1-2) (2003) 89–101.

[283] G.R. Cunha, S.W. Hayward, Y.Z. Wang, Role of stroma in carcinogenesis of the prostate, Differentiation 70 (9) (2002) 473–485.

[284] F.C.W. Wu, A. Tajar, J. Beynon, et al., Identification of late-onset hypogonadism in middle-aged and elderly men, N. Engl. J. Med. 363 (2) (2010) 123–135.

[285] S. Kitahara, K. Yoshida, K. Ishizaka, et al., Stronger suppression of serum testosterone and FSH levels by a synthetic estrogen than by castration or an LH-RH agonist, Endocr. J. 44 (4) (1997) 527–532.

[286] C.N. Robertson, K.M. Robertson, G.M. Padilla, et al., Induction of apoptosis by diethylstilboestrol in hormone-insensitive prostate cancer cell lines, J. Natl. Cancer Inst. 88 (13) (1996) 908–917.

[287] P. Henriksson, O. Edhag, Orchidectomy versus oestrogen for prostatic cancer: cardiovascular effects, Br. Med. J. (Clin-Res-Ed). 293 (6544) (1986) 413–415.

[288] L. Klotz, I. McNeill, N. Fleshner, A phase 1-2 trial of diethylstilbestrol plus low dose warfarin in advanced prostate carcinoma, J. Urol. 161 (1) (1999) 169–172.

[289] P. Iversen, Orchidectomy and oestrogen therapy revisited, Eur. Urol. 34 (Suppl 3) (1998) 7–11.

[290] R. Lundgren, O. Nordle, K. Josefsson, Immediate estrogen or estramustine phosphate therapy versus deferred endocrine treatment in nonmetastatic prostate cancer: a randomized multicenter study with 15 years of followup. The South Sweden Prostate Cancer Study Group, J. Urol. 153 (5) (1995) 1580–1586.

[291] R.E. Langley, D.C. Gilbert, T. Duong, et al., Transdermal oestradiol for androgen suppression in prostate cancer: long-term cardiovascular outcomes from the randomised Prostate Adenocarcinoma Transcutaneous Hormone (PATCH) trial programme, Lancet 397 (10274) (2021) 581–591.

[292] L.B. Bland, M. Garzotto, T.G. DeLoughery, et al., Phase II study of transdermal estradiol in androgen-independent prostate carcinoma, Cancer 103 (4) (2005) 717–723.

[293] A. Mikkola, M. Ruutu, J. Aro, et al., The role of parenteral polyestradiol phosphate in the treatment of advanced prostatic cancer on the threshold of the new millennium, Ann. Chir. Gynaecol. 88 (1) (1999) 18–21.

[294] J.L. Ockrim, E.N. Lalani, M.E. Laniado, et al., Transdermal estradiol therapy for advanced prostate cancer–forward to the past? J. Urol. 169 (5) (2003) 1735–1737.

[295] R.E. Langley, H.G. Kynaston, A.A. Alhasso, et al., A randomised comparison evaluating changes in bone mineral density in advanced prostate cancer: luteinising hormone-releasing hormone agonists versus transdermal oestradiol, Eur. Urol. 69 (6) (2016) 1016–1025.

[296] C. Abate-Shen, M.M. Shen, Mouse models of prostate carcinogenesis, Trends Genet. 18 (5) (2002) S1–S5.

[297] A. Fradet, M. Bouchet, C. Delliaux, et al., Estrogen related receptor alpha in castration-resistant prostate cancer cells promotes tumor progression in bone, Oncotarget 7 (47) (2016) 77071–77086.

[298] I. Takizawa, M.G. Lawrence, P. Balanathan, et al., Estrogen receptor alpha drives proliferation in PTEN-deficient prostate carcinoma by stimulating survival signaling, MYC expression and altering glucose sensitivity, Oncotarget 6 (2) (2015) 604–616, doi:10.18632/oncotarget.2820.

[299] H. Bonkhoff, Estrogen receptor signaling in prostate cancer: Implications for carcinogenesis and tumor progression, Prostate 78 (1) (2018) 2–10.

[300] Z. Weihua, R. Lathe, M. Warner, J.A. Gustafsson, An endocrine pathway in the prostate, ER beta, AR, 5alpha-androstane-3beta,17beta-diol, and CYP7B1, regulates prostate growth, Proc. Natl. Acad. Sci. USA. 99 (21) (2002) 13589–13594.

[301] D. Kastelan, Z. Giljevic, I. Kraljevic, M. Korsic, Selective estrogen receptor modulators: a possible new treatment of osteoporosis in males, Med. Hypotheses 67 (5) (2006) 1052–1053.

[302] W.A. Ricke, S.J. McPherson, J.J. Bianco, et al., Prostatic hormonal carcinogenesis is mediated by in situ estrogen production and estrogen receptor alpha signaling, FASEB J. 22 (5) (2008) 512–520.

[303] K.-M. Lau, K.-F. To, Importance of Estrogenic signaling and its mediated receptors in prostate cancer, Int. J. Mol. Sci. 17 (9) (2016) 1434, doi:10.3390/ijms17091434.

[304] J. Gehrig, S. Kaulfuß, H. Jarry, et al., Prospects of estrogen receptor β activation in the treatment of castration-resistant prostate cancer, Oncotarget 8 (2017) 34971–34979.

[305] E. Di Zazzo, G. Galasso, P. Giovannelli, et al., Estrogens and their receptors in prostate cancer: therapeutic implications, Front. Oncol. (2018). https://doi.org/10.3389/fonc.2018.00002.

[306] G. Halmos, J.M. Arencibia, A.V. Schally, et al., High incidence of receptors for luteinizing hormone-releasing hormone (LHRH) and LHRH receptor gene expression in human prostate cancers, J. Urol. 163 (2) (2000) 623–629.

[307] A.V. Schally, A. Nagy, Chemotherapy targeted to cancers through tumoral hormone receptors, Trends Endocrinol. Metab. 15 (7) (2004) 300–310.

[308] D. Hao, L. Sun, X. Hu, X. Hao, 99m Tc-LHRH in tumor receptor imaging, Oncol. Lett. 14 (1) (2017) 569–578.

[309] D. Margel, A. Peer, Y. Ber, et al., Cardiovascular morbidity in a randomized trial comparing GnRH agonist and GnRH antagonist among patients with advanced prostate cancer and preexisting cardiovascular disease, J. Urol. 202 (6) (2019) 1199–1208.

[310] P. Davey, M.G. Kirby, Cardiovascular risk profiles of GnRH agonists and antagonists: real-world analysis from UK general practice, World J. Urol. 39 (2) (2021) 307–315.

[311] Å. Tivesten, L. Vandenput, F. Labrie, et al., Low serum testosterone and estradiol predict mortality in elderly men, J. Clin. Endocrinol. Metab. 94 (7) (2009) 2482–2488.

[312] G. Corona, G. Rastrelli, G. Di Pasquale, et al., Endogenous testosterone levels and cardiovascular risk: meta-analysis of observational studies, J. Sex. Med. 15 (9) (2018) 1260–1271.

[313] A.T. Porter, B.-J.E. FACRO, Humoral mechanisms in prostate cancer: a role for FSH, Urologic Oncol.: Seminars Original Investigations 6 (4) (2001) 131–138.

[314] P.L. Nguyen, S.M.H. Alibhai, S. Basaria, et al., Adverse effects of androgen deprivation therapy and strategies to mitigate them, Eur. Urol. 67 (5) (2015) 825–836.

[315] A. Kokorovic, A.I. So, H. Serag, et al., Canadian Urological association guideline on androgen deprivation therapy: adverse events and management strategies, Can Urol. Assoc. J. 15 (6) (2021) E307–E322, doi:10.5489/cuaj.7355.

[316] M.F. Townsend, W.H. Sanders, R.O. Northway, G.S.D. Jr, Bone fractures associated with leutinizing hormone-releasing hormone agonists used in the treatment of prostate carcinoma, Cancer 79 (3) (1997) 545–550.

[317] V.B. Shahinian, Y.F. Kuo, J.L. Freeman, et al., Risk of fracture after androgen deprivation for prostate cancer, N. Engl. J. Med. 352 (2) (2005) 154–164.

[318] E.J. Hamilton, A. Ghasem-Zadeh, E. Gianatti, et al., Structural decay of bone microarchitecture in men with prostate cancer treated with androgen deprivation therapy, J. Clin. Endocrinol. Metab. 95 (12) (2010) E456–E463.

[319] S. Schnell, S.M. Friedman, D.A. Mendelson, et al., The 1-year mortality of patients treated in a hip fracture program for elders, Geriatr Orthop Surg Rehavil 1 (1) (2010) 6–14.

[320] J. Compston, A. Cooper, C. Cooper, et al., UK clinical guideline for the prevention and treatment of osteoporosis, Arch Osteoporos 12 (1) (2017) 43.

[321] W. Kemmler, L. Häberle, S. von Stengel, Effects of exercise on fracture reduction in older adults. A systematic review and meta-analysis, Osteoporos. Int. 24 (7) (2013) 1937–1950.

[322] L.D. Gillespie, M.C. Robertson, W.J. Gillespie, et al. Interventions for preventing falls in older people living in the community. Cochrane Database Syst. Rev. 2012(9):CD007146.

[323] F. El-Khoury, B. Cassou, M.A. Charles, P. Dargent-Molina, The effect of fall prevention exercise programmes on fall induced injuries in community dwelling older adults: systematic review and meta-analysis of randomised controlled trials, BMJ 347 (2013) f6234.

[324] B.J. Boucher, Why do so many trials of vitamin D supplementation fail? Endocr. Connect. 9 (9) (2020) R195–R206. https://doi.org/10.1530/EC-20-0274.

[325] V. Nair-Shalliker, A. Bang, S. Egger, et al., Post-treatment levels of plasma 25- and 1,25-dihydroxy vitamin D and mortality in men with aggressive prostate cancer, Sci. Rep. 10 (1) (2020) 7736.

[326] D.R. Taafe, D.A. Galvão, N. Spry, et al., Immediate versus delayed exercise in men initiating androgen deprivation: effects on bone density and soft tissue composition, BJU Int. 123 (2) (2019) 261–269.

[327] R.U. Newton, D.A. Galvão, N. Spry, et al., Exercise mode specificity for preserving spine and hip bone mineral density in prostate cancer patients, Med. Sci. Sports Exerc. 51 (4) (2019) 607–614.

[328] R. Newton, D. Taaffe, D. Galvão, et al., Timing of exercise for muscle strength and physical function in men initiating ADT for prostate cancer, Prostate Cancer Prostatic Dis. 23 (3) (2020) 457–464.

[329] T.J. Martin, V. Grill, Bisphosphonates - mechanisms of action, Aust. Prescr. 23 (2000) 130–132.

[330] M.R. McClung, E.M. Lewiecki, S.B. Cohen, et al., Denosumab in postmenopausal women with low bone mineral density, New England J. Med. 354 (8) (2006) 821–831.

[331] T. Jakob, Y.M. Tesfamariam, S. Macherev, et al., Bisphosphonates or RANK-ligand-inhibitors for men with prostate cancer and bone metastases: a network meta-analysis, Cochrane Database Syst. Rev. 12 (2020) CD013020.

[332] https://arthritisaustralia.com.au/medication-search/bisphosphonates-intravenousiv/https://arthritisaustralia.com.au/medication-search/denosumab.

[333] R.J. Thomas, S.A. Kenfield, A. Jimenez, Exercise-induced biochemical changes and their potential influence on cancer: a scientific review, Br. J. Sports Med. 51 (8) (2017) 640–644.

[334] M.M. Ziaaldini, E. Marzetti, A. Picca, Z. Murlasits, Pathophysiology of sarcopenia and its modulation by exercise. A narrative review, Front. Med. (Lausanne) 4 (2017) 167.

[335] Z. Chen, B. Li, R.-Z. Zhan, et al., Exercise mimetics and JAK inhibition attenuate IFN-γ–induced wasting in engineered human skeletal muscle, Sci. Adv. 7 (4) (2021) eabd9502.

[336] D.A. Galvão, D.R. Taafe, N. Spry, et al., Exercise can prevent and even reverse adverse effects of androgen suppression treatment in men with prostate cancer, Prostate Cancer Prostatic Dis. 10 (4) (2007) 340–346.

[337] D.A. Galvão, D.R. Taafe, N. Spry, et al., Combined resistance and aerobic exercise program reverses muscle loss in men undergoing androgen suppression therapy for prostate cancer without bone metastases: a randomized controlled trial, J. Clin. Oncol. 28 (2) (2010) 340–347.

[338] R.U. Newton, D.A. Galvão, Exercise in prevention and management of cancer, Curr. Treat. Options Oncol. 9 (2-3) (2008) 135–146.

[339] J.-S. Kim, R.L. Wilson, D.R. Taafe, et al., Myokine expression and tumor-suppressive effect of serum following 12 weeks of exercise in prostate cancer patientson ADT, Med. Sci. Sports Exerc. 54 (2) (2022) 197–205.

[340] J.S. Kim, D.A. Galvão, R.U Newton, et al., Exercise-induced myokines and their effect on prostate cancer, Nat. Rev. Urol. 18 (9) (2021) 519–542, doi:10.1038/s41585-021-00476-y.

[341] G. Iolascon, G. Di Pietro, F. Gimigliano, et al., Physical exercise and sarcopenia in older people: position paper of the Italian Society of Orthopaedics and Medicine (OrtoMed), Clin. Cases Miner. Bone Metab. 11 (3) (2014) 215–221.

[342] J.D. Wilson, C. Roehborn, The long-term consequences of castration in men: lessons from the Skoptzy and the eunuchs of the chinese and ottoman courts, J. Clin. Endocrinol. Metab. 84 (12) (1999) 4324–4331.

[343] L. Baumgarten, A.A. Dabaja, Diagnosis and management of gynecomastia for urologists, Curr. Urol. Rep. 19 (7) (2018) 46.

[344] A. Sansone, F. Romanelli, M. Sansone, et al., Gynecomastia and hormones, Endocrine 55 (1) (2017) 37–44.

[345] R.A.S. Davanço, M.S. Neto, É.B. Garcia, et al., Quality of life in the surgical treatment of gynecomastia, Aesthetic Plast. Surg. 33 (4) (2009) 514–517.

[346] A. Heidenreich, P.J. Bastian, J. Bellmunt, et al., EAU guidelines on prostate cancer. Part II: treatment of advanced, relapsing, and castration resistant prostate cancer, Eur. Urol. 65 (2) (2014) 467–479.

[347] M. Oberste, S. Sharma, W. Bloch, P. Zimmer, Acute Exercise-induced set shifting benefits in healthy adults and its moderators: a systematic review and meta-analysis, Front. Psychol. 12 (2021) 528352, doi:10.3389/fpsyg.2021.528352.

[348] R.J. Rohrich, R.Y. Ha, J.M. Kenkel, W.P. Adams Jr., Classification and management of gynecomastia: defining the role of ultrasound-assisted liposuction, Plast. Reconstr. Surg. 111 (2) (2003) 909–923.

[349] G.F. Nicoletti, F. D'Andrea, G. Ferraro, et al., Differential expression of estrogen receptor α and β transcripts in tissues and in primary culture cells from pubertal gynecomastia, Endocrinol Invest. 35 (2) (2012) 150–153.

[350] R. Robeva, A. Elenkova, S. Zacharieva, Causes and metabolic consequences of gynecomastia in adult patients, Int. J. Endocrinol. 2019 (2019) 6718761 eCollection 2019, doi:10.1155/2019/6718761.

[351] R. Pivonello, D. Menafra, E. Riccio, et al., Metabolic disorders and male hypogonadotropic hypogonadism, Front Endocrinol (Lausanne) 25 (10) (2019) 345.

[352] Obesity: preventing and managing the global epidemic. Report of a WHO consultation, World Health Organ. Tech. Rep. Ser. 894 (i-xii) (2000) 1–253.

[353] K. Alberti, R.H. Eckel, S.M. Grundy, et al., Harmonizing the metabolic syndrome: a joint interim statement of the International Diabetes Federation Task Force on Epidemiology and Prevention; National Heart Lunch and Blood Institute; American Heart Association; World Heart Federation; International Atherosclerosis Society; and International Association for the Study of Obesity, Circulation 120 (16) (2009) 1640–1645.

[354] https://www.racgp.org.au/afp/2013/august/the-metabolic-syndrome.

[355] J.M. Dekker, C. Girman, T. Rhodes, et al., Metabolic syndrome and 10-year cardiovascular disease risk in the Hoorn study, Circulation 112 (5) (2005) 666–673.

[356] S. Mottillo, K.B. Filion, J. Genest, et al., The metabolic syndrome and cardiovascular risk: a systematic review and meta-analysis, J. Am. Coll. Cardiol. 56 (14) (2010) 1113–1132.

[357] M. Gomarasca, G. Banfi, G. Lombardi, Myokines: the endocrine coupling of skeletal muscle and bone, Adv. Clin. Chem. 94 (2020) 155–218.

[358] H.S. Chung, K.M. Choi, Adipokines and myokines: a pivotal role in metabolic and cardiovascular disorders, Curr. Med. Chem. 25 (20) (2018) 2401–2415.

[359] P. Gerretsen, B.G. Pollock, Drugs with anticholinergic properties: a current perspective on use and safety, Expert Opin. Drug Saf. 10 (5) (2011) 751–765.

[360] https://www.theseniorlist.com/medication/anticholinergic-drugs/.

[361] T. Mayer, J. Kopitz, K. Plaschke, et al., Limitations of the anticholinergic activity assay and assay-based anticholinergic drug scales, Am. J. Geriatr. Psychiatry 24 (12) (2016) 1182–1188.

[362] A.T. Page, M.O. Falster, M. Litchfield, et al., Polypharmacy among older Australians, 2006–2017: a population- based study, Med. J. Aust. 211 (2) (2019) 71–75.

[363] J. Kluger, A. Roy, H.H. Chao, Androgen deprivation therapy and cognitive function in prostate cancer, Curr. Urol. Rep. 22 (3) (2020) 24–31.

[364] H.J. Green, K.I. Pakenham, B.C. Headley, et al., Altered cognitive function in men treated for prostate cancer with luteinizing hormone-releasing hormone analogues and cyproterone acetate: a randomised controlled trial, BJU Int. 90 (4) (2002) 427–432.

[365] H.J. Green, K.I. Pakenham, B.C. Headley, et al., Quality of life compared during pharmacological treatments and clinical monitoring for non-localised prostate cancer: a randomised controlled trial, BJU Int. 93 (7) (2004) 975–979.

[366] M.M. Cherrier, C.S. Higano, Impact of androgen deprivation therapy on mood, cognition, and risk for AD, Urol. Oncol. 38 (2) (2020) 53−61.

[367] B.D. Gonzalez, H.S. Jim, M. Booth-Jones, et al., Course and predictors of cognitive function in patients with prostate cancer receiving androgen-deprivation therapy: a controlled comparison, J. Clin. Oncol. 33 (18) (2015) 2021–2027.

[368] R.A. Gardiner, J. Yaxley, K.I. Pakenham, et al., Patients receiving androgen deprivation therapy risk adverse cognitive changes, J. Clin. Oncol. 33 (36) (2015) 4314–4315.

[369] A.C. Wilson, M.S. Salamat, R.J. Haasl, et al., Human neurons express type I GnRH receptor and respond to GnRH I by increasing luteinizing hormone expression, J. Endocrinol. 191 (3) (2006) 651–663.

[370] A.V. Schally, Luteinizing hormone-releasing hormone analogs: their impact on the control of tumorigenesis, Peptides 20 (10) (1999) 1247–1262.

[371] B. Dubois, H. Hampel, H.H. Feldman, et al., Preclinical Alzheimer's disease: definition, natural history, and diagnostic criteria, Alzheimers Dement 12 (3) (2016) 292–323.

[372] M. Maggio, E. Dall'Aglio, F. Lauretani, et al., The hormonal pathway to cognitive impairment in older men, J. Nutr. Health Aging 16 (1) (2012) 40–54.

[373] K.T. Nead, G. Gaskin, C. Chester, et al., Androgen deprivation therapy and future Alzheimer's disease risk, J. Clin. Oncol. 34 (6) (2016) 566–571.

[374] K. Richardson, C. Fox, I. Maidment, et al., Anticholinergic drugs and risk of dementia: case-control study, BMJ 361 (2018) k1315, doi:10.1136/bmj.k1315.

[375] T.A. Yap, A.D. Smith, R. Ferraldeschi, et al., Drug discovery in advanced prostate cancer: translating biology into therapy, Nat. Rev. Drug Discov. 15 (10) (2016) 699–718.

[376] K.M. Wadosky, S. Koochekpour, Molecular mechanisms underlying resistance to androgen deprivation therapy in prostate cancer, Oncotarget 7 (39) (2016) 64447–64470.

[377] A.H. Bryce, Y.H. Chen, G. Liu, et al., Patterns of Cancer Progression of Metastatic Hormone-sensitive Prostate Cancer in the ECOG3805 CHAARTED Trial, Eur. Urol. Oncol. 3 (6) (2020) 717–724.

[378] M.A. Villagran, F.A. Gutierrez-Castro, D.F. Pantoja, et al., Bone stroma-derived cells change coregulators recruitment to androgen receptor and decrease cell proliferation in androgen-sensitive and castration-resistant prostate cancer cells, Biochem. Biophys. Res. Commun. 467 (4) (2015) 1039–1045.

[379] T. Visakorpi, E. Hyytinen, P. Koivisto, et al., In vivo amplification of the androgen receptor gene and progression of human prostate cancer, Nat. Genet. 9 (4) (1995) 401–406.

[380] M.J. Linja, K.J. Savinainen, O.R. Saramaki, et al., Amplification and over expression of androgen receptor gene in hormone-refractory prostate cancer, Cancer Res. 61 (9) (2001) 3550–3555.

[381] C.W. Gregory, R.T. Johnson Jr, J.L. Mohler, et al., Androgen receptor stabilization in recurrent prostate cancer is associated with hypersensitivity to low androgen, Cancer Res. 61 (7) (2001) 2892–2898.

[382] B. Gottlieb, H. Lehvaslaiho, L.K. Beitel, et al., The androgen receptor gene mutations database, Nucleic. Acids. Res. 26 (1) (1998) 234–238.

[383] M.E. Taplin, G.J. Bubley, Y.J. Ko, et al., Selection for androgen receptor mutations in prostate cancers treated with androgen antagonist, Cancer Res. 59 (11) (1999) 2511–2515.

[384] A. Sharp, I. Coleman, W. Yuan, et al., Androgen receptor splice variant-7 expression emerges with castration resistance in prostate cancer, J. Clin. Invest. 129 (1) (2019) 192–208.

[385] E.S. Antonarakis, C. Lu, H. Wang, et al., AR-V7 and resistance to enzalutamide and abiraterone in prostate cancer, N. Engl. J. Med. 371 (11) (2014) 1028–1038.

[386] S. Gupta, S. Halabi, G. Kemeny, et al., Circulating tumor cell genomic evolution and hormone therapy outcomes in men with metastatic castration-resistant prostate cancer (mCRPC), Mol. Cancer Res. 19 (6) (2021) 1040–1050, doi:10.1158/1541-7786.MCR-20-0975.

[387] M.T. Schweizer, S. Sivakumar, H. Tukachinsky, et al., Concordance of DNA repair gene mutations in paired primary prostate cancer samples and metastatic tissue or cell-free DNA, JAMA Oncol. 7 (9) (2021) 1–5, doi:10.1001/jamaoncol.2021.2350.

[388] J. Li, M. Berk, M. Alyamani, et al., Hexose-6-phosphate dehydrogenase blockade reverses prostate cancer drug resistance in xenograft models by glucocorticoid inactivation, Sci. Transl. Med. 13 (595) (2021) eabe8226, doi:10.1126/scitranslmed.abe8226.

[389] B. Tombal, F. Saad, D. Penson, et al., Patient-reported outcomes following enzalutamide or placebo in men with non-metastatic, castration-resistant prostate cancer (PROSPER): a multicentre, randomised, double-blind, phase 3 trial, Lancet Oncol. 20 (4) (2019) 556–569.

[390] H.I. Scher, K. Fizazi, F. Saad, et al., Increased survival with enzalutamide in prostate cancer after chemotherapy, N. Engl. J. Med. 367 (13) (2012) 1187–1197.

[391] I.D. Davis, A.J. Martin, M.R. Stockler, et al., Enzalutamide with standard first-line therapy in metastatic prostate cancer, N. Engl. J. Med. 381 (2) (2019) 121–131.

[392] A.J. Armstrong, R.Z. Szmulewitz, D.P. Petrylak, et al., ARCHES: a randomized, phase III study of androgen deprivation therapy with enzalutamide or placebo in men with metastatic hormone-sensitive prostate cancer, J. Clin. Oncol. 37 (32) (2019) 2974–2986.

[393] https://search.tga.gov.au/s/search.html?collection=tga-websites-web&query=enzalutamide+.

[394] T.H. Lin, S.O. Lee, Y. Niu, et al., Differential androgen deprivation therapies with anti-androgens casodex/bicalutamide or MDV3100/enzalutamide versus anti-androgen receptor ASC-J9(R) lead to promotion versus suppression of prostate cancer metastasis, J. Biol. Chem. 288 (27) (2013) 19359–19369, doi:10.1074/jbc.M113.477216.

[395] Y. Niu, C. Guo, S. Wen, et al., ADT with antiandrogens in prostate cancer induces adverse effect of increasing resistance, neuroendocrine differentiation and tumor metastasis, Cancer Lett. 439 (2018) 47–55, doi:10.1016/j.canlet.2018.09.020.

[396] M. Lange, H. Laviec, H. Castel, et al., Impact of new generation hormone-therapy on cognitive function in elderly patients treated for a metastatic prostate cancer: Cog-Pro trial protocol, BMC Cancer 17 (1) (2017) 549 with correction in: Lange M, Laviec H, Castel H et al. BMC Cancer. 2018;18(1):110.

[397] Apalutamide for prostate cancer, Aust. Prescr. 42 (2019) 32–33. https://www.nps.org.au/australian-prescriber/articles/apalutamide-for-prostate-cancer.

[398] K.N. Chi, N. Agarwal, A. Bjartell, et al., Apalutamide for metastatic, castration-sensitive prostate cancer, N. Engl. J. Med. 381 (1) (2019) 13–24.

[399] M.R. Smith, F. Saad, S. Chowdhury, et al., Apalutamide treatment and metastasis-free survival in prostate cancer, N. Engl. J. Med. 378 (15) (2018) 1408–1418.

[400] K.N. Chi, S. Chowdhury, A. Bjartell, et al., Apalutamide in patients with metastatic castration-sensitive prostate cancer: final survival analysis of the randomized, double-blind, phase III TITAN study, J. Clin. Oncol. 39 (20) (2021) 2294–2303.

[401] Apalutamide Prostate cancer UK https://prostatecanceruk.org/prostate-information/treatments/apalutamide.
[402] H. Borgmann, N. Lallous, D. Ozistanbullu, et al., Moving towards precision urologic oncology: targeting enzalutamide-resistant prostate cancer and mutated forms of the androgen receptor using the novel inhibitor darolutamide (ODM-201), Eur. Urol. 73 (1) (2018) 4–8.
[403] A.-M. Moilanen, R. Riikonen, R. Oksala, et al., Discovery of ODM-201, a new-generation androgen receptor inhibitor targeting resistance mechanisms to androgen signaling-directed prostate cancer therapies, Sci. Rep. 5 (2015) 12007, doi:10.1038/srep12007.
[404] E.D. Crawford, W. Stanton, D. Mandair, Darolutamide: an evidenced-based review of its efficacy and safety in the treatment of prostate cancer, Cancer Manag. Res. 12 (2020) 5667–5676.
[405] K. Fizazi, N. Shore, T.L. Tammela, et al., Nonmetastatic, castration-resistant prostate cancer and survival with darolutamide, N. Engl. J. Med. 10;383 (11) (2020) 1040–1049.
[406] https://www.tga.gov.au/sites/default/files/auspar-darolutamide-200702-pi.pdf.
[407] K. Mori, H. Mostafael, B. Pradere, et al., Apalutamide, enzalutamide, and darolutamide for non-metastatic castration-resistant prostate cancer: a systematic review and network meta-analysis, Int. J. Clin. Oncol. 25 (11) (2020) 1892–1900.
[408] C.J. Ryan, M.R. Smith, J.S. de Bono, et al., Abiraterone in metastatic prostate cancer without previous chemotherapy, N. Engl. J. Med. 368 (2) (2013) 138–148.
[409] K. Fizazi, H.I. Scher, A. Molina, et al., Abiraterone acetate for treatment of metastatic castration-resistant prostate cancer: final overall survival analysis of the COU-AA-301 randomised, double-blind, placebo-controlled phase 3 study, Lancet Oncol. 13 (10) (2012) 983–992.
[410] R. Iacovelli, C. Ciccarese, E. Bria, et al., The cardiovascular toxicity of abiraterone and enzalutamide in prostate cancer, Clin Genitourin Cancer 16 (3) (2018) e645–e653.
[411] G. Lu-Yao, N. Nikita, S.W. Keith, et al., Mortality and hospitalization risk following oral androgen signaling inhibitors among men with advanced prostate cancer by pre-existing cardiovascular comorbidities, Eur. Urol. 77 (2) (2020) 158–166.
[412] E.B. Cone, S. Reese, M. Marchese, et al., Cardiovascular toxicities associated with abiraterone compared to enzalutamide-A pharmacovigilance study, E Clinical Medicine 36 (2021) 100887, doi:10.1016/j.eclinm.2021.100887.
[413] H.H. Cheng, R. Gulati, A. Azad, et al., Activity of enzalutamide in men with metastatic castration-resistant prostate cancer is affected by prior treatment with abiraterone and/or docetaxel, Prostate Cancer Prostatic Dis. 18 (2015) 122–127.
[414] F. Saad, E. Efstathiou, G. Attard, et al., Apalutamide plus abiraterone acetate and prednisone versus placebo plus abiraterone and prednisone in metastatic, castration-resistant prostate cancer (ACIS): a randomised, placebo-controlled, double-blind, multinational, phase 3 study, Lancet Oncol. 22 (11) (2021) 1541–1559, doi:10.1016/S1470-2045(21)00402-2.
[415] D.E. Rathkopf, E. Efstathiou, G. Attard, et al. Final results from ACIS, a randomized, placebo-controlled double-blind phase 3 study of apalutamide and abiraterone acetate plus prednisone versus abiraterone in patients (pts) with chemo-naive metastatic castration-resistant prostate cancer. Abstract 9 ASCO. 2021.
[416] L. Marandino, F. Vignani, C. Buttigliero, et al., Evaluation of cognitive function in trials testing new-generation hormonal therapy in patients with prostate cancer: a systematic review, Cancers (Basel) 12 (9) (2020) 2568, doi:10.3390/cancers12092568.

[417] I. Davis, Research Review: Enzalutamide in men with metastatic hormone-sensitive prostate cancer: focus on the ARCHES and ENZAMET trials, Urological Soc. Australia New Zealand invited Monograph (2021). www.researchreview.com.au.

[418] W.T. Lowrance, R.H. Breau, R. Chou, et al., Advanced prostate cancer: AUA/ASTRO/SUO guideline PART I, J. Urol. 205 (1) (2021) 14–21.

[419] C. Parker, E. Castro, K. Fizazi, et al., Prostate cancer: ESMO clinical practice guidelines for diagnosis, treatment and follow-up, Ann. Oncol. 31 (9) (2020) 1119–1134.

[420] National Comprehensive Cancer Network. NCCN Guidelines version 3.2020. Prostate cancer. 2020. 30. Reeves FA, Corcoran N. https://www.bing.com/search?q=National+Comprehensive+Cancer+Network.+NCCN+Guidelines+version+3.2020_prostate&go=Search&qs=ds&form=QBRE.

[421] P.M. Quilty, D. Kirk, J.J. Bolger, et al., A comparison of the palliative effects of strontium-89 and external beam radiotherapy in metastatic prostate cancer, Radiother. Oncol. 31 (1) (1994) 33–40.

[422] A.T. Porter, A.J. McEwan, J.E. Powe, Results of a randomized phase-III trial to evaluate the efficacy of strontium-89 adjuvant to local field external beam irradiation in the management of endocrine resistant metastatic prostate cancer, Int. J. Radiat. Oncol. Biol. Phys. 25 (5) (1993) 805–813.

[423] G.O. Oosterhof, J.T. Roberts, T.M. de Reijke, et al., Strontium(89) chloride versus palliative local field radiotherapy in patients with hormonal escaped prostate cancer: a phase III study of the European Organisation for Research and Treatment of Cancer, Genitourinary Group, Eur. Urol. 44 (5) (2003) 519–526.

[424] C. Collins, J.F. Eary, G. Donaldson, et al., Samarium-153-EDTMP in bone metastases of hormone refractory prostate carcinoma: a phase I/II trial, J. Nucl. Med. 34 (11) (1993) 1839–1844.

[425] A.N. Serafini, S.J. Houston, I. Resche, et al., Palliation of pain associated with metastatic bone cancer using samarium-153 lexidronam: a double-blind placebo-controlled clinical trial, J. Clin. Oncol. 16 (4) (1998) 1574–1581.

[426] S. Dizdarevic, R. McCready, S. Vinjamuri, Radium-223 dichloride in prostate cancer: proof of principle for the use of targeted alpha treatment in clinical practice, Eur. J. Nucl. Med. Mol. Imaging 47 (1) (2020) 192–217.

[427] C.C. Parker, R.E. Coleman, O. Sartor, et al., Three-year Safety of Radium-223 Dichloride in Patients with Castration-resistant Prostate Cancer and Symptomatic Bone Metastases from Phase 3 Randomized Alpharadin in Symptomatic Prostate Cancer Trial, Eur. Urol. 73 (3) (2018) 427–435.

[428] A. Hijab, S. Curcean, N. Tunariu, et al., Fracture risk in men with metastatic prostate cancer treated with radium-223, Clin Genitourin Cancer 19 (5) (2021) e229–e305, doi:10.1016/j.clgc.2021.03.020.

[429] R.S. Israeli, C.T. Powell, W.R. Fair, W.D. Heston, Molecular cloning of a complementary DNA encoding a prostate-specific membrane antigen, Cancer Res. 53 (2) (1993) 227–230.

[430] R.S. Israeli, C.T. Powell, J.G. Corr, et al., Expression of the prostate-specific membrane antigen, Cancer Res. 54 (7) (1994) 1807–1811.

[431] G.L. Wright, B.M. Grob, C. Haley, et al., Upregulation of prostate-specific membrane antigen after androgen-deprivation therapy, Urology 48 (2) (1996) 326–334.

[432] S.D. Sweat, A. Pacelli, G.P. Murphy, D.G. Bostwick, Prostate-specific membrane antigen expression is greatest in prostate adenocarcinoma and lymph node metastases, Urology 52 (4) (1998) 637–640.

[433] J.K. Troyer, M.L. Beckett, G.L. Wright Jr., Detection and characterization of the prostate-specific membrane antigen (PSMA) in tissue extracts and body fluids, Int. J. Cancer 62 (5) (1995) 552–558.

[434] M.S. Hofman, J. Violet, R.J. Hicks, et al., [^{177}Lu]-PSMA-617 radionuclide treatment in patients with metastatic castration-resistant prostate cancer (LuPSMA trial): a single-centre, single-arm, phase 2 study, Lancet Oncol. 19 (2018) 825–833.

[435] M.J. Roberts, A. Morton, P. Donato, et al., (68)Ga-PSMA PET/CT tumour intensity pre-operatively predicts adverse pathological outcomes and progression-free survival in localised prostate cancer, Eur. J. Nucl. Med. Mol. Imaging 48 (2) (2021) 477–482.

[436] O. Sartor, J. de Bono, K.N. Chi, et al., Lutetium-177–PSMA-617 for metastatic castration-resistant prostate cancer, N. Engl. J. Med. 385 (12) (2021) 1091–1103.

[437] B.A. Nock, A. Kaloudi, E. Lymperis, et al., Theranostic perspectives in prostate cancer with the gastrin-releasing peptide receptor antagonist NeoBOMB1: preclinical and first clinical results, J. Nucl. Med. 58 (1) (2017) 75–80.

[438] M.S. Hofman, L. Emmett, S. Sandhu, et al., [^{177}Lu]Lu-PSMA-617 versus cabazitaxel in patients with metastatic castration-resistant prostate cancer (TheraP): a randomised, open-label, phase 2 trial, Lancet 397 (309276) (2021) 797–804.

[439] D. Meyrick, M. Gallyamov, S. Sabarimurugan, et al., Real-world data analysis of efficacy and survival after Lutetium-177 labelled PSMA ligand therapy in metastatic castration-resistant prostate cancer, Target. Oncol. 16 (3) (2021) 369–380.

[440] O. Sartor, J. de Bono, K.N. Chi, et al., Lutetium-177-PSMA-617 for metastatic castration-resistant prostate cancer, N. Engl. J. Med. 385 (12) (2021) 1091–1103, doi:10.1056/NEJMoa2107322.

[441] T. Kitamura, Necessity of re-evaluation of estramustine phosphate sodium (EMP) as a treatment option for first-line monotherapy in advanced prostate cancer, Int. J. Urol. 8 (2) (2001) 33–36.

[442] V. Patel, B. Liaw, W. Oh, The role of ketoconazole in current prostate cancer care, Nat. Rev. Urol. 15 (10) (2018) 643–651.

[443] I.F. Tannock, D. Osoba, M.R. Stockler, et al., Chemotherapy with mitoxantrone plus prednisone or prednisone alone for symptomatic hormone-resistant prostate cancer: a Canadian randomized trial with palliative end points, J. Clin. Oncol. 14 (6) (1996) 1756–1764.

[444] I.F. Tannock, R. de Wit, W.R. Berry, et al., Docetaxel plus prednisone or mitoxantrone plus prednisone for advanced prostate cancer, N. Engl. J. Med. 351 (14) (2004) 1502–1512.

[445] D.P. Petrylak, C.M. Tangen, M.H. Hussain, et al., Docetaxel and estramustine compared with mitoxantrone and prednisone for advanced refractory prostate cancer, N. Engl. J. Med. 351 (14) (2004) 1513–1520.

[446] C.J. Sweeney, Y.-H. Chen, M. Carducci, et al., Chemohormonal therapy in hormone-sensitive prostate cancer, N. Engl. J. Med. 373 (8) (2015) 737–746.

[447] N.D. James, M.R. Sydes, N.W. Clarke, et al., Addition of docetaxel, zoledronic acid, or both to first-line hormone therapy in prostate cancer (STAMPEDE): survival results from an adaptive, multiarm, multistage, platform, Lancet 387 (10024) (2016) 1163–1177.

[448] Y. Shan, Y. Huang, A.M. Lee, et al., A long noncoding RNA, GAS5 can be a biomarker for docetaxel response in castration resistant prostate cancer, Front. Oncol. 11 (2021) 675215, doi:10.3389/fonc.2021.675215.

[449] S.D. Fossa, N.K. Aaronson, D. Newling, et al., Advanced hormone resistant prostate cancer: preliminary observation on subjective morbidity and palliation, Eur. Urol. 18 (1990) 2 abstract.

[450] V.J. Sinibaldi, Docetaxel treatment in the elderly patient with hormone refractory prostate cancer, Clin. Interv. Aging. 2 (4) (2007) 555–560.

[451] E. Aotani, T. Hamano, A. Gemma, et al., Identification of adverse events that have a negative impact on quality of life in a clinical trial comparing docetaxel versus S-1 with cisplatin in lung cancer, Int. J. Clin. Oncol. 21 (5) (2016) 836–842.

[452] M. Lange, N. Heutte, O. Rigal, et al., Decline in cognitive function in older adults with early-stage breast cancer after adjuvant treatment, Oncologist 21 (11) (2016) 1337–1348.

[453] H.L. Rush, L. Murphy, A.K. Morgans, et al. Quality of life in men with prostate cancer randomly allocated to receive docetaxel of abiraterone in the STANPEDE trial. 2021;JCO2100728. doi: 10.1200/JCO/21.00728.

[454] A.A. Azad, N. Dhiantravan, L. Emmett, et al., UpFrontPSMA: a randomized phase 2 study of sequential 177Lu-PSMA617 and docetaxel vs docetaxel in metastatic hormone-naïve prostate cancer (clinical trial protocol), BJU Int. 128 (3) (2021) 331–342.

[455] D.H. Kwon, J. Chou, S.M. Yip, et al., Differential treatment outcomes in BRCA1/2-, CDK12-, and ATM-mutated metastatic castration-resistant prostate cancer, Cancer 127 (12) (2021) 1965–1973, doi:10.1002/cncr.33487.

[456] J.M. Sperger, H. Emamekhoo, R.R. McKay, et al., Prospective evaluation of clinical outcomes using a multiplex liquid biopsy targeting diverse resistance mechanisms in metastatic prostate cancer, J. Clin. Oncol. 39 (26) (2021) 2926–2937.

[457] K.A. Menear, C. Adcock, R. Boulter, et al., 4-[3-(4-cyclopropanecarbonylpiperazine-1-carbonyl)-4-fluorobenzyl]-2H-phthalazin-1-one: a novel bioavailable inhibitor of poly(ADP-ribose) polymerase-1, J. Med. Chem. 51 (20) (2008) 6581–6591.

[458] J. De Bono, J. Mateo, K. Fizazi, et al., Olaparib for metastatic castration-resistant prostate cancer, N. Engl. J. Med. 382 (22) (2020) 2091–2102.

[459] S. Loeb, V.N. Giri, Clinical implications of germline testing in newly diagnosed prostate cancer, Eur. Urol. Oncol. 4 (1) (2021) 1–9.

[460] J.S. de Bono, N. Mehra, G.V. Scagliotti, et al., Talazoparib monotherapy in metastatic castration-resistant prostate cancer with DNA repair alterations (TALAPRO-1): An open-label, phase 2 trial, Lancet Oncol. 22 (9) (2021) 1250–1264.

[461] https://medlineplus.gov/druginfo/meds/a618070.html.

[462] C. Sweeney, S. Bracarda, C.N. Sternberg, et al., Ipatasertib plus abiraterone and prednisolone in metastatic castration-resistant prostate cancer (IPATential150): a multicentre, randomised, double-blind, phase 3 trial, Lancet 398 (10295) (2021) 131–142.

[463] M. Markowski, R.H. Getzenberg, M.S. Steiner, et al., The effect of VERU-111, a novel oral inhibitor of α and β tubulin, on tumor growth in the human castration-resistant, AR-variant prostate cancer (PCa) model 22Rv1, J. Clin. Oncol. 37 (2019) 167.

[464] M.C. Markowski, R.F. Tutrone, M.A. Eisenberger, et al., VERU-111, an oral cytoskeleton disruptor, to treat men with metastatic castration-resistant prostate cancer (mCRPC) who failed an androgen receptor targeting agent, J. Clin. Oncol. 39 (15 suppl) (2021) 5056.

[465] W.B. Coley, The treatment of inoperable sarcoma by bacterial toxins (the mixed toxins of the streptococcus erysipelas and the bacillus prodigiosus), Proc. R. Soc. Med. 3 (Surg Sect) (1910) 1–48, doi:10.1177/003591571000301601.

[466] Y. Jin, A. Tan, J. Feng, et al., Prognostic impact of memory CD8(+) T cells on immunotherapy in human cancers: a systematic review and meta-analysis, Front. Oncol. 11 (2021) 698076, doi:10.3389/fonc.2021.698076.

[467] P.D. Reid, A.S. Cifu, A.R. Bass, Management of immunotherapy-related toxicities in patients treated with immune checkpoint inhibitor therapy, JAMA 325 (5) (2021) 482–483, doi:10.1001/jama.2020.17308.

[468] T. Kubo, T. Shinkawa, Y. Kikuchi, et al., Fundamentals and essential knowledge for pathologists engaged in the research and practice of immune checkpoint inhibitor-based cancer immunotherapy, Front. Oncol. 11 (2021) 679095, doi:10.3389/fonc.2021.679095.

[469] J. Larkin, V. Chiarion-Sileni, R. Gonzalez, et al., Combined nivolumab and ipilimumab or monotherapy in untreated melanoma, N. Engl. J. Med. 373 (1) (2015) 23–34, doi:10.1056/NEJMoa1504030.

[470] E.S. Antonarakis, J.M. Piulats, M. Gross-Goupil, et al., Pembrolizumab for treatment-refractory metastatic castration-resistant prostate cancer: multicohort, open-label Phase II KEYNOTE-199 study, J. Clin. Oncol. 38 (5) (2020) 395–405.

[471] W. Abida, M.L. Cheng, J. Armenia, et al., Analysis of the prevalence of microsatellite instability in prostate cancer and response to immune checkpoint blockade, JAMA Oncol. 5 (4) (2019) 471–478.

[472] Z.R. Reichert, J. Urrutia, J.J. Alumkal, et al., Microsatellite instability as an emerging biomarker for checkpoint inhibitor response in advanced prostate cancer, JAMA Oncol. 5 (2019) 478–479.

[473] P. Barata, N. Agarwal, R. Nussenzveig, et al., Clinical activity of pembrolizumab in metastatic prostate cancer with microsatellite instability high (MSI-H) detected by circulating tumor DNA, J. Immunother. Cancer 8 (2) (2020) e001065.

[474] A. Ribas, J.D. Wolchok, Cancer immunotherapy using checkpoint blockade, Science 359 (2018) 1350–1355.

[475] E.D. Kwon, C.G. Drake, H.I. Scher, et al., Ipilimumab versus placebo after radiotherapy in patients with metastatic castration-resistant prostate cancer that had progressed after docetaxel chemotherapy (CA184-043): a multicentre, randomised, double-blind, phase 3 trial, Lancet Oncol. 15 (7) (2014) 700–712.

[476] P. Sharma, R.K. Pachynski, V. Narayan, et al., Nivolumab plus ipilimumab for metastatic castration-resistant prostate cancer: preliminary analysis of patients in the checkmate 650 trial, Cancer Cell 38 (4) (2020) 489–499.e3.

[477] S.J. Priceman, E.A. Gerdts, D. Tilakawardane, et al., Co-stimulatory signaling determines tumor antigen sensitivity and persistence of CAR T cells targeting PSCA+ metastatic prostate cancer, Oncoimmunology 7 (2) (2017) e1380764.

[478] R.P. Junghans, Q. Ma, R. Tathore, et al., Phase I Trial of anti-PSMA designer CAR-T cells in prostate cancer: possible role for interacting interleukin 2-T cell pharmacodynamics as a determinant of clinical response, Prostate. 76 (14) (2016) 1257–1270.

[479] G.P. Murphy, B.A. Tjoa, H. Ragde, et al., Phase I clinical trial: T-cell therapy for prostate cancer using autologous dendritic cells pulsed with HLA-A0201-specific peptides from prostate-specific membrane antigen, Prostate 29 (6) (1996) 371–380.

[480] R.A. Gardiner, C. Schmidt, K.A.O. Ellem, et al., The Northern Section, USANZ. A randomised controlled study of a DC-based vaccine study in patients with hormone escaping prostate cancer, BJU Int. 101 (2008) 1 (Abstract 29 p9).

[481] R.A. Gardiner, C. Schmidt, K.A.O. Ellem, et al., Northern Section of Urological Society of Australia & New Zealand. A phase I study with metastatic tumour as the antigen source in patients with advanced prostate cancer, BJU Int. 103 (2009) 1 (Abstract 39 p12).

[482] R.L. Prue, F. Vari, K.J. Radford, et al., A phase I clinical trial of CD1c (BDCA-1)+ dendritic cells pulsed with HLA-A*0201 peptides for immunotherapy of metastatic hormone refractory prostate cancer, J. Immunother. 38 (2) (2015) 71–76.

[483] P.W. Kantoff, C.S. Higano, N.D. Shore, et al., Sipuleucel-T immunotherapy for castration-resistant prostate cancer, N. Engl. J. Med. 363 (2010) 411–422.

[484] B. Mastelic-Gavillet, A. Sarivalasis, L.E. Lozano, et al., Quantitative and qualitative impairments in dendritic cell subsets of patients with ovarian or prostate cancer, Eur. J. Cancer 135 (2020) 173–182.

[485] E.P. Pinkham, L. Teleni, J.L. Nixon, et al., Conventional supportive cancer care services in Australia: a national service mapping study (The CIA study), Asia Pac. J. Clin. Oncol. 18 (3) (2022) 191–200 Online ahead of print, doi:10.1111/ajco.13575.

[486] G.G. Taets, C.M. Fernandes, L. Torres, C.M. Taets, Biological effects of music in cancer patients: contributing to an evidence -based practice, Int. J. Sciences 9 (04) (2020) 21–27.

[487] B. Koczwara, R. Meng, M.D. Miller, et al., Late mortality in people with cancer: a population-based Australian study, Med. J. Aust. 214 (7) (2020) 318–323, doi:10.5694/mja2.50879.

[488] S. Chowdhury, D. Robinson, D. Cahill, et al., Causes of death in men with prostate cancer: an analysis of 50 000 men from the Thames Cancer Registry, BJU Int. 112 (2013) 182–189, doi:10.1111/bju.12212.

[489] G.N. Levine, A.V. D'Amico, P. Berger, et al., Androgen-deprivation therapy in prostate cancer and cardio-vascular risk: A science advisory from the American Heart Association, American Cancer Society, and American Urological Association: endorsed by the American Society for Radiation Oncology, Circulation 121 (2010) 833–840. https://doi.org/10.1161/CIRCULATIONAHA.109.192695.

[490] U.S. Food & Drug Administration. FDA drug safety communication: update to ongoing safety review of GnRH agonists and notification to manufacturers of GnRH agonists to add new safety information to labeling regarding increased risk of diabetes and certain cardiovascular diseases. 2010. https://www.fda.gov/Drugs/DrugSafety/ucm229986.htm.

[491] S.J. Freedland, P.-A. Abrahamsson. Androgen deprivation therapy and side effects: are GnRH antagonists safer? Asian J. Androl. 23 (1) (2021) 3–10.

[492] S. Dai, Y. Mo, Y. Wang, et al., Chronic stress promotes cancer development, Front. Oncol. 10 (2020) 1492, doi:10.3389/fonc.2020.01492.

[493] H. Mohammadpour, C.R. MacDonald, P.L. McCarthy, et al., $\beta 2$-adrenergic receptor signaling regulates metabolic pathways critical to myeloid-derived suppressor cell function within the TME, Cell Rep. 37 (4) (2021) 109883.

[494] J. Cuzick, M.A. Thorat, G. Andriole, et al., Prevention and early detection of prostate cancer, Lancet Oncol. 15 (11) (2014) e484–e492, doi:10.1016/S1470-2045(14)70211-6.

[495] D. Sarfati, B. Koczwara, C. Jackson, The impact of comorbidity on cancer and its treatment, CA Cancer J. Clin. 66 (2016) 337–350.

[496] B.K. Edwards, A.M. Noone, A.B. Mariotto, et al., Annual Report to the Nation on the status of cancer, 1975–2010, featuring prevalence of comorbidity and impact on survival among persons with lung, colorectal, breast, or prostate cancer, Cancer 120 (9) (2014) 1290–1314.

[497] R.I. Jordan, M. Allsop, Y. El Mokhallalati, et al., Duration of palliative care before death in international routine practice: a systematic review and meta-analysis, BMC Med. 18 (1) (2020) 368.

[498] https://www.who.int/health-topics/palliative-care.

[499] https://www.nice.org.uk/guidance/csg4.

[500] J.-S. Kim, D.A. Galvão, R.U. Newton, et al., Exercise-induced myokines and their effect on prostate cancer, Nat. Rev. Urol. 18 (9) (2021) 519–542, doi:10.1038/s41585-021-00476-y.

[501] H. Contreras-Trujillo, J. Eerdeng, S. Akre, et al., Deciphering intratumoral heterogeneity using integrated clonal tracking and single-cell transcriptome analyses, Nat. Commun. 12 (1) (2021) 6522, doi:10.1038/s41467-021-26771-1.

Addendum

The following is a list of information and support organisations that is far from comprehensive but may be of help. Readers from countries other than those mentioned below may find the following informative **National and International Cancer Organisations | CancerIndex** www.cancerindex.org/clinks7.htm and **Cancer Resources by Country – Asia and Africa – CancerIndex** www.cancerindex.org/clinks5a.htm.

Asia

Korean Cancer Association https://www.uicc.org cancer@cancer.or.kr https://www.cancer.or.kr telephone 00 82 2 792 1486.
Japanese Cancer Association (JCA) http://www.jca.gr.jp/english/index.html.
00 81 3 5361 7156 office@jca.gr.jp
Philippine Society of Medical Oncology provides an information on patient support programs and groups, agencies for government and non-government financial assistance https://psmo.org ph (02) 721-9326 or (02) 723-0101 loc 5318.
Singapore Cancer Society https://portal.singaporecancersociety.org.sg 1800 727 3333 provides information and support.
Taiwan Cancer Total-Care Information Society Contact information Society http://www.tci-mandarin.com/ 00886225115101 tci@totalcare.org.tw.
The Prostate Cancer Society Malaysia (PCSM) pcsmmalaysia@gmail.com.
The Prostate Cancer Society Malaysia (PCSM) is a network of support groups across Malaysia run by prostate cancer survivors voluntarily.
National Cancer Society Malaysia is a charity helping patients with cancer and their families through education, care, and through support groups mhelp@cancer.org.my Toll free 1-800-88-1000.

Australia

The **Cancer Councils** are a network of state and territory-based charities in a federation, with *Cancer Council Australia* the overarching central body. Cancer Councils (e.g., Cancer Council Queensland) provide a wide range of services to support patients and their families with prostate cancer one of the major conditions of their focus. For the cost of a local telephone call from anywhere in Australia, the *Cancer Council Helpline* provides a free, confidential support service with specially trained staff available to answer questions and provide support.

Prostate Cancer Foundation of Australia is a community organisation and the peak national body for prostate cancer in Australia dedicated to reducing the impact of prostate cancer, partners, and families. www.pcfa.org.au/support or freecall 1800 220 099.

Both the above organisations work collaboratively with **Beyond Blue** https://www.beyondblue.org.au and **Healthy Male** https://www.healthymale.org.au in coordinating advice and support services.

Brazil

Cancer Society of Brazil Sociedade Brasileira de Cancerologia www.sbcancer.org.br.

Canada

The **Canadian Cancer Society** is a body of volunteers and staff with expertise in research advocacy, cancer control, information, and support through a wide range of services. https://cancer.ca/en/about-us/prostate-cancer.

In addition, the following websites may be useful sources of information.

Prostate cancer – Canada.ca

Prostate Cancer Canada Network

Prostate Cancer Research Institute

Movember – Men's Health – Prostate Cancer

Europe

Cancer Support France is a network of independent associations throughout France that provides support for English-speaking people of any nationality who have been touched by cancer. Each association organises itself according to the needs of its own area and all associations are set. https://cancersupportfrance.org.

Europa Uomo The European Prostate Cancer Coalition is an advocacy group that aims to support men with prostate cancer improving quality of life. The organisation represents 27 national patients' groups across Europe. The Central Office is in Antwerp Belgium Tel: +32 3 645 94 44 | Fax: | Email: europauomo@skynet.be https://www.europa-uoomo.org.

New Zealand

Prostate Cancer Foundation of New Zealand https://prostate.org.nz has a range of resources for the public that include Ministry of Health Guidelines, prostate cancer and diet, managing prostate cancer, conferences, and fora together with other services.

Cancer Society NZ https://www.cancer.org.nz provides a range of support services that include a range of resources for prostate cancer patients and their families. Support groups across New Zealand are coordinated mainly by trained volunteers, and meet monthly or bimonthly, providing opportunities for discussion, and input from professionals involved with prostate cancer patients.

Prostate cancer- Health Navigator New Zealand https://www.healthnavigator.org.nz/health-a-z/p/prostate-cancer.

Prostate cancer registry -PCOR-ANZ https://prostatecancerregistry.org.

South Africa

A list of South African Cancer Organisations (including prostate cancer) is available at www.cancerindex.org/South_Africa Included amongst these are the following.

Prostate Cancer Foundation of South Africa is a nonprofit organisation promoting public awareness, education, advocacy, and providing support. https://prostate-ca.co.za Prostate Cancer Helpline 084 283 3982.

The Cancer Association of South Africa is a community-driven and volunteer-based organisation with 45 regional offices. The association promotes health in all communities within South Africa, through advocacy and supporting research, prevention, early detection, and care. CANSA also provides Support through support groups, online support resources, tele counseling amongst other services. https://cansa.org.za 0800 22 6622 072 197 9305 English and Afrikaans 071 867 3530 Xhosa, Zulu, Sotho and Siswat.

United Kingdom

Prostate Cancer UK https://prostatecanceruk.org is an organisation with support groups throughout much of the UK. A free hotline and live website are run by specialist nurses accessible via the website or by telephoning 0800 074 8383.

Prostate Scotland www.prostatescotland.org.uk is a charity providing information and help on prostate cancer with support groups, exercise programs, workshops, and other services across Scotland.

Tackle prostate cancer https://tackleprostate.org is a UK-wide charity representing people with prostate cancer and those who care for them. The organisation has a free national helpline on 0800 035 5302 or email helpline@tackleprostate.org available for discussing any prostate cancer issue and is able to connect to support groups locally for those who contact them.

PCaSO (Prostate Cancer South Support Charity) is an organisation for men diagnosed with prostate cancers run by volunteer prostate cancer patients working from home. In addition to information and telephone contact numbers for discussing prostate cancer issues accessed via the website, there are regular meetings in Dorset, Hampshire, and Sussex https://pcaso.org.
CHAPS The Men's Health Charity provides information on a number of men's health issues including prostate cancer via their general website https://www.pinterest.com.au/chapshealth/ or via chaps.uk.com/prostate-cancer.

United States of America

Us Too is a support, education, and advocacy non-profit organisation that provides comprehensive educational materials and resources along with support services to more than 200 volunteer-led support groups across the United States and abroad. https://ustoo.org US Too can be contacted by telephone (630) 795-1002, email ustoo@ustoo.org or via their helpline (800) 80-Us TOO (800-808-7866). Their office hours are 9 am–5 pm Central Time.
Prostate Cancer Foundation https:www.pcf.org/contact Although the Prostate Cancer Foundation (PCF) is prominent as a research foundation, it also provides educational information including a register of clinical trials as well as information on support groups and financial assistance.
MaleCare is a men's cancer survivor support and advocacy national non-profit organisation, run by oncologists, psychologists, and social workers. Their goal is to help men and their loved ones live longer and happier lives with services provided without cost to patients and their families. MaleCare's services extend to provide programs that focus on survivorship for gay and bisexual men as well as African-American men. https://malecare.org/about/.
The American Cancer Society provides a 24/7 helpline (1-800-227-2345), education, transport, and lodges as well as supporting research and providing advocacy.
https://www.cancer.org/treatment/support-programs-and-services.
Prostate Cancer International provides a wide spectrum of services and programs that are primarily intended for prostate cancer patients themselves, men at particular levels of risk for prostate cancer, men who wish to understand and manage their risk for prostate cancer, and their family members. https://pcainternational.org.
The Prostate Net is a nonprofit patient education and advocacy organisation dedicated to providing an interactive online network to maximise broad-scale, mass communications of actionable health messages for Prostate Cancer Victims and their caregivers. https://www.prostatenet.org.

Index

Page numbers followed by "*f*" and "*t*" indicate, figures and tables respectively.